Engendering the Energy System

Joy Clancy

Engendering the Energy System

Looking Back to Go Forward

palgrave
macmillan

Joy Clancy
75inQ
Rotterdam, The Netherlands

ISBN 978-3-031-78964-9 ISBN 978-3-031-78965-6 (eBook)
https://doi.org/10.1007/978-3-031-78965-6

This Palgrave Macmillan imprint is published by the registered company Springer Nature Switzerland AG
The registered company address is: Gewerbestrasse 11, 6330 Cham, Switzerland

This book is dedicated to
ENERGIA
The international network on gender and energy

Acknowledgements

This book contributes to the quest to leave no-one behind in the Green Energy Transition which as a researcher I have been involved in for more than 40 years. I started my career as a scientist working on technical aspects of renewable energy systems. However, the intended beneficiaries did not share my enthusiasm for what I saw as the solution to their problems. I asked: why not? What's the problem? This became a light bulb moment in which I began to recognise that proposing a technocratic response without understanding the context is not going to work. While I have not lost my enthusiasm for renewable energy technologies, a major part of my research has focused on the social dimension of the energy system. I am proud to say that I have the distinction of being one of the early researchers in the field of gender and energy.

In this book I have given an overview of the evolution of gender and energy in research and practice and our current understanding of the gender perspective in the energy system. I have also made some suggestions about where we need further research and how to proceed. Many people have contributed to that understanding. First, I would like to thank the colleagues who reviewed the chapters: Elizabeth Cecelski, Andrew Barnet, Marielle Feenstra, Ana Rojas, Debajit Palit, Amanda Elan, Josephine Musango and Stefan Bouzarovski. Their critical comments are much appreciated. I would also like to thank the two anonymous reviewers of the proposal for the helpful suggestions. The book is the better for their input.

In writing this book, I have drawn on the research projects I have been involved with which have provided valuable data and experiences through field trips which gave me insight into the contexts I was trying to understand. I would particularly like to mention ENERGIA and the User TCP Gender and Energy Research Programme which have provided enjoyable collaborative events, through which I have met some amazing people who have contributed to my evolution as a gender and energy researcher. Through collaboration I have learned the benefits of multi-disciplinary research providing more fine-grained insights and understanding on the nature and role of technology in generating a socially just green energy transition from a gender and intersectional perspective. Of course, I cannot forget my students who have also taught me a lot! They have also contributed to the evidence I cite.

I should also acknowledge an informal group known as 'the Energy Ladies' which emerged when we edited a book together: Marielle Feentra, Nthabi Mohlakoana, Lillian Sol Cueva, Gül Őzerol and Leila Niamir. Our collaboration continues through a WhatsApp group through which we share good and bad news, let off steam and provide inspiration and ideas. Sisterhood at its best. Long may it continue.

I am indebted to all the women and men who have contributed to research (my own and others) by providing insights, experiences, opinions and aspirations related to their lived experiences and how the green energy transition can contribute to those aspirations. Without their willingness to share, research findings to identify problems and propose solutions would be the poorer and less appropriate. I trust that I can do them justice.

Last but not least I have to thank my editors Uma Vinesh and Rachael Ballard who have endured endless delays and missed deadlines. I trust they will feel the wait has been worthwhile.

Contents

List of Figures

List of Tables

List of Boxes

CHAPTER 1

Introduction

1.1 Analysing the Gender and Energy Nexus

The assumption on which this book is based is that gender, the social construction of roles and relations between women and men, and the energy system are inextricably linked. The actions in one area have consequences for or impacts in the other, in other words there is a nexus[1] between gender and energy. For some readers this may be a surprising statement. The influence that the energy system has on practical issues in other areas such as water and food is easy to see—for example, energy is used to pump water and water is used to generate electricity. Conversely, how the social system influences the energy system, beyond people requiring energy to do something, or the social relations between women and men are in turn influenced by the energy system is more difficult to recognise. However, these mutual influences are what this book sets out to explore.

The energy system consists of a number of chains based on a particular resource (e.g. coal, oil, sun, wind, biomass) which are supplied to a consumer in a usable form to meet a need (light, heat, entertainment) or provide a service (run a pump or power a sewing machine). Each

[1] A nexus is a concept to describe the situation in the economy where two or more elements, or sectors, are inextricably intermeshed and that actions in one area have impacts in one or more of the others (IEA and the World Bank 2015).

J. Clancy, *Engendering the Energy System*,
https://doi.org/10.1007/978-3-031-78965-6_1

chain has several stages: extraction, production, conversion, transportation within or across national boundaries, delivery and use. At each stage there are several elements that are coherently organised and interconnected to deliver energy in a usable form: hardware (tools and machines), organisation of supply (centralised/decentralised; stand-alone; formal/informal) and people who play a range of roles in the design, organisation and operation of the chains as well as being customers. Each chain has its own structure with specific organisations, institutions, regulations, cultures and practices. While it is tempting to visualise the energy system as formed from physical structures, such as power stations, wind turbines and solar panels, we should not lose sight of the people involved in the system. People are not only consumers but also work for energy companies, for the government as policymakers and regulators and for civil society organisations advocating for clean energy. As individuals and as employees, they bring to the system knowledge and skills as well as their interests, norms, values and standards of behaviour and good practice (Geels 2004).

The way a chain is organised and run can be changed, as well as which chains are prioritised by the government. In other words, it is a political decision (Meadowcroft 2011; van de Graaf and Sovacool 2020). For example, liberalisation of energy markets in the 1980s saw a move away from energy companies under state ownership to the private sector being the major player. Nevertheless, governments still play a key role in the energy sector through legislative frameworks related to energy pricing and subsidy reform, as well as the energy resources which will make up the chains of the energy system.

The elements of a chain and the way in which they are organised and interact can change, for example, in terms of user practices, industrial networks and infrastructure. These transformation processes, in which existing structures are broken down and new ones are established can take one or two generations to materialise. A range of events can initiate the transformation. For example, the oil crisis of the 1970s was fundamentally a political action by producing countries that undermined the energy security of recipient countries. As a consequence, an interest was stimulated in exploiting indigenous renewable energy sources aimed at reducing a dependence on the political agendas of producing countries. More recently environmental issues linked to climate change and fossil fuel use have initiated the transformation to a low-carbon society. Here the transformation is more radical than the changes in the 1970s since it is not only new technologies and a diversification of the energy sources at the start of the chain but also new forms of organisation of supply are envisaged. Liberalised energy markets have allowed other actors to enter the system including decentralised options owned and operated by a range of actors, including communities and households.

However, these transformations in the energy sector have not had universal benefits. As of 2021, an estimated 675 million people had little or no access to electricity while 2.3 billion did not have access to clean cooking (fuels and technology) (IEA et al. 2023)—in other words they are living in energy poverty—a condition in which people do not have access to sufficient quantities of modern energy to meet their daily needs (Sovacool 2012). There are several consequences of living in energy poverty, such as ill-health and limited capacity to earn a living. It may come as a surprise to some readers that a considerable number of these people are living in the Global North. For example, it is estimated that, in 2023, in the European Union around 31 million people (nearly 7% of the population) have difficulty paying their energy bills or have limited access to high-quality energy and hence live in energy poverty (Widuto 2023). However, when these data are unpacked gender and other social issues influencing who lives in energy poverty begin to emerge. For example, globally women suffer more severely than men from energy poverty (Clancy et al. 2017; UNIDO-UN Women 2013).

Note on Terms Used to Categorise Countries

It is part of research to categorise what is being studied. The terms used to label the categories can be contested. In the context of development, I use the terms 'North' and 'South' to locate countries rather than the terms 'developing' and 'developed'. These terms are not used in the geographical sense but in their socio-economic and political sense. Indeed, one can argue that all countries are 'developing' since the word implies change and all countries are constantly changing economically, politically and socially albeit at different rates at different points in time. The 'North' (sometimes referred to as the Global North) includes the United States, Canada, Europe, and parts of Asia with significant industrial base (e.g. Japan), as well as Australia and New Zealand. The 'South' (or the Global South) is made up of Africa, Latin America, Caribbean and Pacific Islands and less industrialised parts of Asia including the Middle East. At some point in the last 100 years, many of the countries included in the South have been colonies, which has significantly influenced the basis of their economies as well as their social and political institutions. Countries in the North at the aggregate level, based on some measure such as GDP/capita, are economically wealthier than those countries in the South. However, there is a distinct blurring of these lines between countries and within countries. There are pockets of 'the South' in 'the North' and vice versa. *Over the last 20 years there have begun to emerge from the South a group of countries (which includes Brazil, Russia, India and China known as the BRIC countries) in which their economies have moved from agriculturally based to industrial production. This illustrates the problem of trying to categorise a moving target!*

When looking to explain these numbers and the apparent failure of the energy system, we need to remember the statement above that the energy transition is a political decision. The transition brings change not only to the physical components but also the human components, potentially disrupting established political interests which will be resisted (Meadowcroft 2011).

1.2 Energy Access Through a Gender Lens

Gender issues in the energy sector can be analysed at the macro, meso- and micro-levels of society. The macro-level is the policy arena, which has its own levels of operation, that is, national, regional and international. The meso-level is where formal and informal organisations and institutions operate. The micro-level is the space an individual occupies, such as the household, community or workplace. The three levels are inter-connected.

At the *macro-level,* policymakers have assumed in energy policy debates and when formulating legislative frameworks related to energy pricing, subsidy reform, rural energy policy and energy technology that the outcomes benefit women and men equally, in other words they are gender-neutral. However, it has been argued that energy policies are gender-blind. In the formulation of policy, there is a failure to recognise the differences in the energy needs of women and men together with the difference in assets to meet those needs. Gender blindness leads to policy failures, for example, a continued reliance on traditional biomass fuels for cooking with their negative health consequences (see Chapter 3). Such policy failures are rooted in gender relations which are the social arrangements governing who does what, who gets to consume what and who takes what decisions. Gender relations are shaped by the balance of power between men and women functioning in a combination of cooperation and conflict at all levels in society. Gender relations, like gender roles, are socially determined. Hence, they are influenced by the same social, cultural, political and economic institutions both informal, such as in the family, and formal, such as in the legal system. These institutions govern who has the power to make decisions about all aspects of daily life—in general men have more power than women to make decisions, at all levels in society, not only about their lives but also the lives of others. These gendered institutional barriers are the basis of the gender blindness of the energy system since women have limited influence over energy policy formulation and implementation (Clancy and Feenstra 2006).

At the meso-level, in terms of energy organisations delivering modern energy services, two important gender issues have been the focus of attention. Firstly, how to ensure that gender is incorporated into the operation of energy utilities both in terms of energy access and in terms of opening up employment opportunities for women beyond administration and finance. Secondly, how to change mindsets that see women as passive users and consumers of energy into ones that see that these experiences can also be utilised by women as energy entrepreneurs across a range of energy services and technologies. Women know local circumstances and understand local needs so they should play a key role in determining policy decisions. A woman may be able to sell more effectively to other women, and access to potential female clients is not hindered by social constraints.

At the micro-level, the gender division of labour leads to different demands for quantities and forms of energy. In the first instance, gender equality would appear to be satisfied by providing the right quantities and types of energy with the appropriate technologies for women and men to meet their daily needs. In reality achieving equitable outcomes through energy interventions is more complex than merely supplying a piece of equipment. How energy access and choices are determined within the household is governed by the gendered institutional barriers referred to above (Moser 1993). The limited decision-making for women includes the acquisition of energy end-use technologies. Women tend to be more disadvantaged than men in similar circumstances; for example, women's access and control over assets such as land, cash and credit is usually more limited than men's. Women's technical skills are often less than men's; for example, compared to men, in many places women's reading levels are lower and they have less experience with hardware. Women in general are 'time-poor' and tend to work a longer day than men do and in salaried occupations, globally, women tend to earn less than men, even in like for like jobs, which can influence their capacity to save for old age. This means that energy interventions aimed to help those with low incomes are likely to benefit men differently from women, in part due to their different capacities to respond.

1.3 Aims of the Book

This book examines the evidence in respect of the transformations at the nexus of the energy system and gender roles and relations. Energy, unlike other infrastructure-related sectors such as water, transport and ICT, has also not been a central topic within the social sciences, including anthropology. There is therefore no extensive empirical literature available with a social science perspective on gender and energy. There are very few studies longitudinal studies observing the social changes caused by an energy intervention. Unfortunately, much of the evaluation literature related to energy projects, including the stoves literature, does not take a gendered perspective, often referring instead to actors as 'people' or 'consumers'. There is also another part of the literature which uses 'women' and 'gender' interchangeably.

I have mainly drawn on findings based on research carried out using recognised scientific methods by independent observers using a range of methods and frameworks. The data sets are mainly qualitative. Unfortunately, there are very few gender and energy studies using ethnographic approaches which require the researchers to embed themselves into a community for extended periods of time, allowing for more detailed interaction with community members, for observations and opportunities to explore and explain. Such approaches offer different insights from standard questionnaire surveys.

The lack of gender disaggregation in statistical studies limits the usefulness of these correlations in determining differential impacts on women and men, or girls and boys; and particularly in assessing their relationship with gender equity and empowerment. The failure to disaggregate data also hides other differences, for example, between rich and poor, urban and rural, and between generations.

Nevertheless, as we will see in Chapter 2, the topic of gender and energy has attracted interest for over 40 years. So, I chose the title of the book to reflect my own evolution in thinking about and experience of research related to the social dimension of the energy system in those decades. The book looks at how our thinking has changed over time, to see what lessons we have learned and what does or doesn't work so we can move forward to ensuring we have an energy system that is socially just and gender-transformative. When attention was first given to the social dimensions of the energy system, the focus was women in the rural areas of the Global South, with cooking as the core issue—men's energy needs were not mentioned. However, the concept of gender began to emerge initially as a binary but increasingly disaggregated across a range of social characteristics at least within research-development practice and policies makers still have a way to go. We also moved outside of the household and into the world of work and how energy access would enable women to earn a living, again men were not visible. Around the start of the twenty-first century, the Global North began to take an interest in gender and energy, although, just as in the Global South, the initial focus was on women.

The book begins with a review of the evolution in thinking about gender in energy (Chapter 2). It then moves to looking at the way in which access to modern energy influences women's and men's daily lived experiences (Chapter 3). Chapter 4 examines how the energy system influences women's capacity to earn an income. Energy policy provides the enabling environment for any transformation, so Chapter 5 looks at the extent to which energy policy is gender-responsive. The final chapter is a reflection on how gender and energy research has developed since the 1980s, which could be considered its birth, what we have learned and what still needs to be done.

References

Clancy, J., Daskalova, V., Feenstra, M., Franceschelli, N. and Sanz, M. (2017). *Gender Perspective on Access to Energy in the EU*, Study for the FEMM Committee of the EU Parliament, PE 596.816, Brussels, 2017.

Clancy, J.S. and Feenstra, M. (2006). *How to Engender Energy Policy*. Technical Paper Prepared for ENERGIA.

Geels, F.W. (2004). From Sectoral Systems of Innovation to Socio-technical Systems: Insights About Dynamics and Change From Sociology and Institutional Theory. *Research Policy* 33 (6–7): 897–920.

IEA, IRENA, UNSD, World Bank and WHO. (2023). *Tracking SDG 7: The Energy Progress Report*. Washington, DC: World Bank.

International Energy Agency (IEA) and the World Bank. (2015). *Sustainable Energy for All 2015—Progress Toward Sustainable Energy*. Washington, DC: World Bank. https://doi.org/10.1596/978-1-4648-0690-2

Meadowcroft, J. (2011). Engaging with the Politics of Sustainability Transitions. *Environmental Innovation and Societal Transitions* 1 (1): 70–75.

Moser, C. (1993). *Gender Planning and Development: Theory, Practice and Training*. London: Routledge.

Sovacool, B. (2012). Energy Security: Challenges and Needs. *WIREs Energy Environ* 1: 51–59. https://doi.org/10.1002/wene.13

UNIDO-UN Women (United Nations Industrial Development Organization-UN Women). (2013). *Sustainable Energy for All: The Gender Dimensions*. New York: UN Women.

Van der Graaf, T. and Sovacool, B.K. (2020). *Global Energy Politics*. Cambridge, UK: Polity Press.

Widuto, A. (2023). *Energy Poverty in the EU*. European Parliamentary Research Service.

CHAPTER 2

Evolution of Thinking on Gender and Energy

2.1 Introduction

Thinking on gender and energy has taken time to emerge as a nexus with a difference in focus between the Global South and the Global North. Gender and energy, or more accurately, women and energy appear as a subject of interest in the development discourse early in the 1980s. In the Global North the issue takes off at the start of the new millennium, although there were some early pioneers again from a perspective of women, rather than gender, and energy. However, in this chapter we start a little earlier by first tracing the evolution of the way energy has been treated in the development discourse to address poverty and how the concept of gender has entered this discourse with a particular focus at the global level within the United Nations (UN) system. The chapter starts with the description of the evolutionary path at the beginning of the 1970s because of two events which are significant in the development of the gender and energy nexus. The first event is related to energy and the second event is related to the publication of a book. The two sections in this chapter on the gender and energy nexus in the development discourse and the Global North give only a very brief overview of the historical development of the approaches in the two contexts, up to around the start of the new millennium. Chapters 3 to 5 go into more detail about more recent advances in gender and energy related to users, employment and policy.

J. Clancy, *Engendering the Energy System*,
https://doi.org/10.1007/978-3-031-78965-6_2

2.2 Energy in the Development Discourse

The end of the Second World War, which is the point in time that this chapter takes as its starting point, brought many changes, including the establishment of international organisations such as the United Nations (UN). One of the tasks facing the UN at the end of the 1940s was to address the inequality in wealth between and within countries. The answer, at least in the dominant economic thinking of the time, was to take as a model the historical experiences of the European countries and the United States (US) which involved a transformation from agriculturally based economies to industrial ones out of which would come improvements in the quality of life. Transformation was required not only in the economic system but also in the social and political systems, in other words, societies were to become modern, that is acquiring the characteristics of industrialised countries, through a process of development.

One of the factors that contributed to industrialisation was energy: the substitution of human and animal labour by fossil fuels (Stern 2011). At the macro-level, it has long been accepted that there is a strong relationship between energy and economic growth which has shaped energy policy (IDS 2003). Energy policies focused on the supply side expanding production capacity which required large quantities of financial investment. Electricity, in particular, was identified with modernity and was the energy form which received the financial support from donors, particularly for rural areas, while countries with coal and gas reserves set about exploiting them.

In the early 1970s, much of the world's attention was focused on oil. The price rises associated with the Arab–Israeli Six Day War drew attention not only to the economic consequences of high oil prices but also the vulnerability of economies to fuel imports. In addition, it was a timely reminder that at an unspecified and uncertain date in the future oil would run out. The consequences included industrialised countries looking to improving energy efficiency and an interest emerged in developing alternatives to fossil fuels (Lee and Usman 2018). The latter was also driven by an increased awareness of the impact burning fossil fuels was having on the environment. For the rural areas of the South, access to electricity was proving problematic due to the cost of grid extension to dispersed populations and in many cases having to cross difficult geographical terrain. Decentralised energy production based on renewable resources was known to be technically possible, however, some argued that a political change in thinking was required to help speed up access to modern energy based on decentralised production. Up to this point in time, energy supply was in many countries a state monopoly. However, the 1970s saw the emergence of policies, based on a neo-liberal political agenda, which aimed to reduce the role of the state in the provision of goods and services, including energy, and for the private sector to take on this role while the state's role was to set and enforce regulations. For countries in the South, a requirement for an energy sector loan from an international financial institution, such as the World Bank, was to open the sector to the market and allowing new actors to be energy suppliers (Lee and Usman 2018). There was an unexpected benefit for rural areas where large energy companies were reluctant to provide energy supplies since they considered the market too small to repay the investment costs of widely dispersed rural populations. Non-state actors less motivated by profits, such as NGOs and community organisations, could also use this new policy framework to become energy providers, aided by developments in decentralised renewable energy technologies bringing costs down and increasing the possibility for rural households to have access to a modern energy supply. What rural households would use this energy for brought a shift in analysis to the demand side of end-uses and the support system required, such as consumer credit, to promote access (Martinot et al. 2002). It is not only electricity that not-for profit organisations have been enabling access, for example, in India, the Odisha Rural Development and Marketing Society (ORMAS) has been working with women's self-help groups to promote clean cooking using LPG, for at least 38,000 households (Kelkar et al. 2019). The Africa Biogas Partnership Programme had (by 2018) brought a clean cooking fuel to more than 320,000 people in five Sub-Saharan countries.[1]

Although policy was driven by electrification, there were initiatives aimed to improve supply of petroleum fuels to rural areas. For example, in India, the national government created several initiatives to increase access to LPG by rural households particularly in more remote locations and those with a potentially small market. A first step, in 2009, was a

[1] https://snv.org/project/africa-biogas-partnership-programme-abpp (accessed 29 August 2022).

programme[2] to enable the private sector to establish small-scale LPG distribution companies (Sinha 2012). In 2014, the central government provided support towards the cost of getting a connection which was considered a barrier to many low-income households having access to a clean cooking fuel. Government data indicates that, as of April 2018, there were more than 35 million new connections added since the introduction of the new subsidy (PPAC 2018 cited in IISD et al. 2019), although these new connections may be due to other factors. Nevertheless, a study of the LPG supply chain in 10 countries in the Global South found an inefficient delivery system, with evidence of unsafe handling of LPG cylinders and commercial malpractice (Kojima 2011). South Africa introduced a system of registered installers and accredited inspectors to increase safety and reduce tampering with the gas composition (Kojima 2011).

Data from the International Energy Agency (IEA), for electrification rates as of 2015 show that for many countries in the South, despite involving the private sector, universal electricity access is still not a reality which is leading to a reassessment of the policies that have shaped the energy sector since the 1980s (Foster and Rana 2020).

2.3 Energy and Poverty

It took until the 1990s before any significant attention was paid at the micro-level to the effects that economic growth has had on poverty which at this point was still assessed in terms of levels of income. The development discourse had for some time considered that people had to have sufficient income to provide themselves with basic needs, although there was no consensus about what these 'basic needs' consisted of with the possible exceptions of food and water. Despite the role energy plays in cooking and boiling water, energy was never considered a basic need (Ramani and Heijndermans 2003). Within the research community a search began looking for evidence that increased energy use could have an impact on poverty. Indications were that energy can play in combating poverty through: (i) improved health; (ii) increased productivity and new opportunities for additional income; (iii) reduced labour and time spent on household activities (see, for example, World Bank 1996, 2000;

[2] Rajiv Gandhi Grameen Vidyutikaran Yojana (RGLV) Pariyojana.

UNDP 2006). These categories are linked. For example, in rural areas, increases in household income could be reached by increasing output through mechanising agricultural production using diesel engines and electricity. Mechanisation can also bring additional benefits: it can reduce drudgery as well as save time which can be used for income generation or rest and recuperation. As early as 1995, it was possible to show a positive link between energy consumption and the Human Development Index (HDI), which is a composite measure of development used within the UN system (UNDP/UNIES 2000). The link between increased HDI and increase in energy consumption is not linear. The greatest impact is found for low-income countries with low levels of modern energy consumption where an increase in modern energy consumption produces a larger, more rapid increase in HDI scores than for countries at higher levels in the HDI (Modi 2004). However, it is difficult empirically to attribute measurable poverty impacts solely to improved energy access since there are many other contributing factors operating simultaneously (IDS 2003).

During the 1990s, within the development discourse there started to evolve a conceptualisation of poverty beyond an economic one related to household income to a broader one encompassing a range of basic goods and services, such as lack of access to adequate levels of food, water, clothing, shelter, sanitation, health care and education (Reddy 2000). This conceptualisation was taken a step further by a landmark study published by the World Bank: 'Can Anyone Hear Us? Voices From 47 Countries' (Narayan 1999). The study showed that people categorised as poor are not a homogeneous group. They vary in extent of their poverty as well as what causes their poverty which was found to have a distinct gender dimension. For adult women, a change in civil status through divorce, widowhood or desertion are common causes of living in poverty, while for men, unemployment and illness (which can be linked) are commonly cited reasons. Again, the routes out of poverty differ for women and men due to their different assets. Women tend to be more disadvantaged than men living in similar circumstances; for example, compared to men, women have more limited access and control over assets such as land, cash and credit. Women have in general less influence over decisions and exercise less control over their own lives and resources—both at the household and community levels—than men (Moser 1993). Women's technical skills are often less than men's, which will influence their opportunities for earning an income. This difference in skills has its roots in the socialisation processes of girls and boys in which the activities and subjects are considered suitable for girls and boys are governed by social norms. Boys are encouraged to explore the physical world and how things work, which contributes to girls having less experience with hardware and a reluctance to engage with technology (UNESCO 2017). This means that energy interventions aimed to help the poor are likely to benefit men differently from women. Research about poverty also showed that there were differences based on whether you live in rural or urban areas, remoteness being particularly significant, and which ethnic group you belong to (Reddy 2000). When trying to move people of poverty, an understanding of the multifaceted nature of poverty would enable the formulation of targeted policies which points to a need to disaggregate data across different social identities.

Another shift in the development discourse during the 1990s could be detected in which energy was seen as a service rather than as a product changing the perception of the way energy contributes to moving people out of poverty (see, for example, World Bank 1996). Evaluations of energy projects and with a better understanding of poverty began to draw attention to some of the assumptions about the outcomes of improved energy access. The World Bank had found that when rural areas were electrified, it was usually the wealthier households which benefitted from electricity access, not the poor (IDS 2003). This was also found to be case with LPG (Kojima 2011).

The United Nations, a major agenda setter for influencing the issues to be addressed by the global community, paid little direct attention to energy during the 1980s and 1990s. The focus was on environmental issues linked to sustainable development culminating in 1992 at the United Nations Conference on Environment and Development. This conference, which has come to be more popularly known by the short form of host city's name (Rio), produced a document (known as Agenda 21) that shaped the development path the global community was to follow for a decade. There is no specific chapter in the document on energy although it is acknowledged as a cross-cutting issue as well as a major cause of climate change. As part of the processes leading on from Agenda 21, world leaders agreed in 2000 to a set of global targets known as the Millennium Development Goals (MDGs) to be reached by 2015. The goals address a range of issues including to halve extreme poverty, improving health and education access and gender equality. Despite the evidence about the role energy plays in all the goals, there is no specific MDG dedicated to energy. It is only in 2002, at the World Summit on Sustainable Development (WSSD), held to review global achievements since the 1980s, that the global development community began to recognise more explicitly the role of energy in sustainable development including in achieving the MDGs.

A more explicit recognition of the role of energy in development can be seen in the successor to the MDGs, the Sustainable Development Goals (SDGs) which have a specific energy goal (SDG7): to ensure access to affordable, reliable, sustainable and modern energy for all. This is framed as a technical, hardware issue. However, the SDGs are intended to be mutually reinforcing which underlines the link between energy access and gender equality (SDG5). The mutually reinforcing aspect of the SDGs could be seen as a return to the approach of integrated rural development from the 1960s in which projects often included a major energy component recognising the key role energy plays in delivering other outcomes. Integrated rural development was abandoned in the 1980s when a sectoral approach was adopted by the international development agencies.

The energy SDG is closely aligned with another UN programme Sustainable Energy for All (SEforALL) which was established in 2011 as an initiative of the UN Secretary General. SEforALL has three goals: ensure universal access to modern energy services, double the global rate, of improvement in energy efficiency, and double the share of renewable energy in the global energy mix. Two main issues are driving the need for a transformation of the energy system: climate change, linked to fossil fuel use, and the lack of access by a substantial portion of the world's population to modern energy sources. When SEforALL was established, it was estimated that 1.1 billion people had no access to electricity, while 2.9 billion had to cook with polluting, inefficient fuels such as firewood. This sorry situation related to energy access is more than 40 years after the World Resources Institute drew our attention to 'the other energy crisis' (Eckholm 1975).

2.3.1 Energy Poverty

During the 1990s the micro-level became the focus of attention within the energy and development discourse, and new ideas and concepts began to emerge. Researchers and practice began to talk about access to energy services[3] rather than energy sources since it is the output, such as heat, light, entertainment, which people want rather than the energy source (Woroniuk and Schalkwyk 1998). This change in formulating the problem from one of energy as a product to the perspective of what people need energy for creates a shift from the supply side to the demand side, requiring an understanding of the constraints or conditions around these end uses. Such considerations underlines the point that energy use is for meeting more ends than economic production—there are also social outcomes such as health and education which require energy inputs. These outcomes also have implications for economic production. The energy user is concerned not only about the price they have to pay for their energy supply but additional characteristics such as accessibility (particularly for those living in remote areas), adequacy, quality, reliability and safety (Reddy 2000). An analysis of the user and the nature of their energy use reveals a distinct link with poverty. Wealthier households purchase high-quality fuels, which meet the demand for a greater variety of end-uses than will be enjoyed by poor households who tend to rely on poor-quality fuels usually based on biomass. Location is a factor in energy choices. Nearly all rural households, irrespective of income, use biomass, which is collected for free, with expenditure limited to lighting sources such as candles and kerosene. Urban households generally buy their fuels which can include biomass fuels such as charcoal (ESMAP 1999). Low-income urban households are found to spend as much as 20% of their income on fuels which is a higher proportion of their income used for this purpose than used by higher income households (Barnes 1995). Low-income households in urban areas also find sources of 'free' wood, such as timber yards, discarded packaging and the urban hinterland (Leach and Mearns 1988). Urban size, based on population, has also been found to be a factor in wood fuel use. Government data for Pakistan in the early 1980s reported that 93% of households in towns with a population of less than 25,000 depended on wood fuels for cooking and heating, which decreased to 65 to 75% in medium-sized cities and in the large cities (more than 250,000 inhabitants) the number fell to 25% (Leach and Mearns 1988).

[3] Energy services can be defined as the desired and useful products, processes or services that result from the use of energy; for example, illumination, comfortable indoor climate, refrigerated storage, transportation and cooking (Annecke 1999).

One outcome of limited energy access for poor households is that less water is boiled for drinking and other hygiene purposes, increasing the likelihood of water-borne diseases. Illness, which can be caused or exacerbated by the fuels they use, reduces the ability of poor people to improve their livelihoods, making them more vulnerable, and exacerbating their poverty. Poor households struggle to buy the equipment needed to use modern energy. These differences in energy use between wealthy and poor households show that there is an energy dimension of poverty which has become conceptualised as 'energy poverty'. There is no universally agreed definition of energy poverty although a commonly cited definition is: the absence of sufficient choice in accessing adequate, affordable, reliable, high-quality, safe and environmentally benign energy services to support economic and human development (Reddy 2000). A person can be considered living in energy-poverty if the energy services of their home does not allow them to participate in the lifestyles, customs and activities which define membership of the society where they live (Buzar 2007). Energy poverty interacts with other manifestations of poverty. However, energy poverty should not only be equated with income poverty. Research from India suggests that the extent of energy poverty can exceed that of income poverty. In rural areas, wealthier households can also be living in energy poverty where a lack of infrastructure and markets creates significant access barriers to electricity, clean-combusting fuels and equipment (Pachauri and Spreng 2011). Also, in the Global North, high-income households can use woodstoves for space heating which users justify as a contribution to environmental protection since biomass is seen as a renewable energy source (Bouzarovski and Petrova 2015). If the measure of energy poverty is taken as a specified percentage of household income allocated to energy bills (for example 10%), there can be households which fall below the percentage but do not have a level of energy services that meets their physical needs and promotes social inclusion. Such households adopt coping strategies, such as cooling/heating only certain rooms in the house, changing their sleeping hours and limiting visits from family members (Stojilovska et al. 2021).

Since the concept of energy poverty emerged, a lot of effort has gone into developing a definition with a wide acceptance as well as getting agreement on what constitutes a basic basket of energy services. Agreement on definitions allows for the development of a metric which enables tracking progress with cross-country comparisons and the success or failure of energy interventions. From a gender perspective what matters is that, irrespective of which measure is used, the data are presented in a way that reveals the gender dimensions and their social differentiation. In that way gender equitable policies can be developed and targeted.

The household energy service which has received most attention is cooking with those living in energy poverty being reliant on wood stoves for the delivery of this service. So, for cooking the transition from energy poverty was seen as a linear progression from traditional smoky biomass fuels, usually obtained for no financial cost, to purchasing modern fuels and their associated stoves. A differentiation was made in the nature of modern fuels, which was price and efficiency related, which would influence household behaviour in fuel choices. It was assumed that households would first switch to relatively cheap sources, such as kerosene, and finally to more expensive but cleaner energy forms, like LPG and electricity. The transition was envisaged in the form of an energy ladder with traditional fuels on the bottom rung, with the cheaper modern fuels on the next rung and the more expensive, but more efficient, fuels at the top. Status of acquiring a new stove and its fuel as a visual indicator of wealth was considered to be a driver for moving up the ladder (Barnes and Floor 1996). However, real life is more complicated than as portrayed in simple economic models. Research showed that other factors than status drive the transition (Masera et al. 2000). Indeed, gender is an important factor—an issue we will return in Chapter 3.

Access to modern energy forms alone is not sufficient to move people out of energy poverty. A reliable supply of the energy source can be a starting point, particularly for helping with income generation which in turn can help pay for energy and the equipment to provide the desired services (Bensch et al. 2012; Rao 2013). Access to electricity has been shown to bring new enterprises, and hence new employment opportunities, to rural areas. For example, in Tanzania, metalworking, bicycle repair and mineral processing activities became available after electrification (Pachauri et al. 2013). However, more factors of input are required to start a new business than simply having access to electricity. People need access to finance to buy equipment that can use modern energy sources. The sums of money required might be beyond the assets of families who are the usual sources of finance for many entrepreneurs and farmers. Access to formal lenders may be constrained in terms of both their availability in rural areas and their willingness to lend without levels of assets, such as land, which may be beyond the hopeful borrower. Women are generally more disadvantaged than men in terms of access to cash and credit with their lack of ownership of land acting as a particular constraint in accessing credit (Narayan 1999). Low level of skills in utilising new equipment and business development, as well as poor access to markets, have also been identified as barriers to income generation (Legros et al. 2011).

2.4 From Women to Gender in the Development Discourse

The incorporation of gender into the development discourse has passed through three phases as summarised in Fig. 2.1: women empowerment; gender mainstreaming and social inclusion. This section describes these three phases starting with the landmark publication in 1970 by Ester Boserup which is recognised as helping bring women as actors in their own right into the development discourse. Boserup disputed the argument prevalent at the time in economic development thinking that it was only necessary to train men to use new technologies since the benefits of their use, such as increased agricultural production and incomes, would automatically reach women (Boserup 1970). She argued that by denying women access to these new opportunities reduced their chances of employment. Boserup's work had an immediate impact. In the United States (US), women involved in development initiatives and galvanised by the Women's Movement, used Boserup's findings to persuade legislators to pass into law the requirement for gender-sensitive impact studies for all development projects to promote gender equality through equal opportunities (Parpart et al. 2000).

The concept of 'Women in Development' (WID) emerged within the development discourse and into development practice. A recognition, at least by the UN system, of women's part in development was the Decade for Women initiated in 1975. WID supporters advocated for improved women's access to education, training, property ownership and credit as well as their incorporation into the market economy which would contribute to economic development.[4] The response from donors and development agencies was to set up WID offices and units. The way that WID was translated into practice was not without its critics. It has been

[4] There is empirical evidence to support the notion that gender equality has a positive impact on economic growth, although for the reverse, the evidence is mixed (Kabeer 2012a).

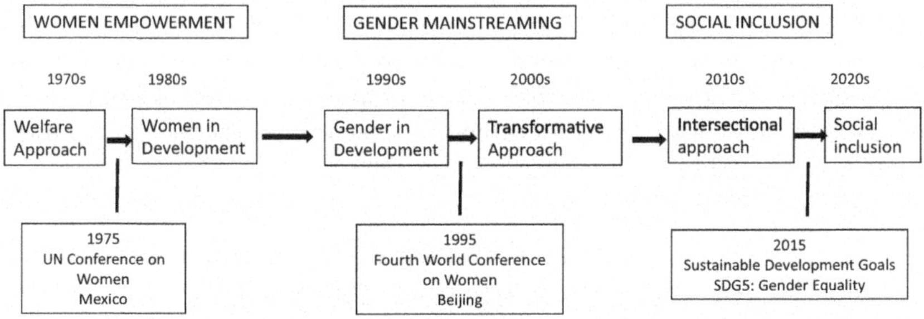

Fig. 2.1 Historical development of gender integration into the development discourse (Feenstra 2021: 40)

seen as a top-down, technocratic, donor-driven approach with a focus on providing statistical measures of women's work and an emphasis on employment (Parpart et al. 2000). Feminists considered that the focus on employment devalues women's work within the household and their contribution to the care economy (Listo 2018). On the other hand, WID as a concept was considered to have helped create images of women as 'mothers', 'carers' and 'holders of tradition and culture' which form a rather simplistic picture of the complexity of women's lives devoid of context (Escobar 1995). WID advocates considered that it was women's primary role in care work and responsibility for household chores which helped explain their exclusion from productive activities. Women from the South considered that the approach treated women as a homogeneous group and did not take into account issues related to race and class, as well as the global inequalities between the South and the North. A more general criticism was a lack of attention to women's subordination, failing to address the underlying causes of why and how women are marginalised. From a research perspective, WID can be credited with recognising that women and men had different experiences of development and so enabling the separate study of women's experiences and perceptions within the economy from those of men (Rathgeber 1990).

A group of radical feminists in academia and activism, particularly from the South, critical of the WID approach called for women's projects to be separate from men's. This 'separatist' approach to development became known as Women and Development (WAD). Those who identified with WAD argued that women's subordination is rooted in patriarchy which, they argued, if women are to benefit from development, then a different approach is required to that espoused by WID which was embedding women into a patriarchal world of production. WAD was also critical of WID's emphasis on individuals and their capacity for self-betterment, ignoring opportunities for collective action and support against resistance by men (and some women) to any changes in gender relations and the distribution of benefits accruing from development (Young 1993). There was also criticism by activists and feminists in the South that Northerners, including feminists, present women in the South as passive victims instead of 'agents of change'. Perhaps the most important contribution by those arguing for a WAD approach has been in drawing attention to women's distinctive knowledge, work, goals and responsibilities. Also, WAD pointed out that women had always been part of development, for example, in agriculture, yet their contributions were systematically overlooked by national and international organisations (Parpart et al. 2000).

The WAD approach was also criticised. WADs origins are located in Marxists critiques of capitalism which treats women as a class and so neglects differences between women such as ethnicity (Scott 1995). The ideology of patriarchy received little attention. The problems of social relations between women and men are subsidiary to the inequalities and oppression of the international economic and political system experienced by women and men. Improvements in women's lives, it was argued, will only come with changes in the international system (Rathgeber 1990). From an outcome perspective, the promotion of women-only projects, as part of the WAD approach, was criticised as being small-scale and underfunded, in part because they tended to have a grassroots focus implemented by non-governmental organisations (NGOs) and community-based organisations (CBOs) which it is claimed has limited the transformative potential of initiatives across scale[5] (Jones et al. 2015).

Dissatisfaction with both WID and WAD led to the emergence of a third approach: Gender and Development (GAD). The GAD approach brings together positive elements of WID and WAD while addressing the criticisms. GAD addresses patriarchy and its maintenance of women in a subordinate position, as well as addressing women's material conditions and position in all levels of the economy and society which are reflected in their social status. GAD distances itself from women-only projects since

[5] Women do organise themselves collectively at scale, for example, the Self-Employed Women's Association (SEWA) in India.

these do not specifically deal with gender relations which, as these relations are socially constructed, require the structures and processes that lead to women's subordination to change. GAD advocates saw the role of women's groups as a means to increase women's political power rather than the WID approach of enabling women's productive capacity. A GAD approach recognises that women are not a homogeneous group but differ according to a range of characteristics and experiences such as race, class and colonial history. Women's inequalities are also perpetuated by global inequalities. A significant difference between GAD and WID/WAD is the recognition by the former that men who are concerned about issues of equity and social justice can also play a role in gender equality (Sen and Grown 1987). As with the WID approach, there have been strong criticisms of GAD from the South. Indeed, a group of feminists from the South established a network Development Alternatives for Women in a New Era (DAWN) to challenge the dominance of the North in the development debate (Tambe and Trotz 2010). There is a tendency to overlook women in the South's subordination as part of a colonial and post-colonial legacy. For example, in colonial systems which introduced private land ownership, women more often than men lost land-tenure rights (Sen and Grown 1987). Feminists from the South have also criticised the homogenisation of women from the South who are portrayed as subordinate to their male partners—a view which ignores variations in the geographic and contextual locations which shape gender relations.

Meanwhile events within the UN system led to the second phase of women in the development discourse with the emergence of another concept: gender mainstreaming (see Fig. 2.1). In 1995, the Fourth World Conference on Women (held in Beijing, China, which is the name by which the Conference is frequently referred to particularly in relation to the document produced at the end of the Conference—the Beijing Declaration), held as part of the follow-up processes leading on from the UN Decade for Women mentioned above, drew global attention to the unequal opportunities sometimes faced by women and men for a myriad of reasons including differences in legal status and property rights. As a contribution to the Conference, UNDP introduced a 'gender-related development index' (GDI) to reflect gender disparities at the country level. The conference also stressed that women are not passive 'receptors' of development but that women can and do play a role in contributing to solutions including the design and use of technology. To achieve gender equality, the conference concluded that gender had to be brought centre stage into the development process, in other words gender was to be brought from the margins into the mainstream. In future, gender equality concerns would be integrated into the analyses and formulation of all UN policies, programmes and projects. Gender mainstreaming would enable women as well as men to influence the development agenda (Moser 2005). Women would be 'empowered' to participate in processes. Post-Beijing, there was the establishment of gender desks/departments in many ministries and development agencies as well as several policy initiatives considered as successful steps towards gender equality, such as the introduction of legal reforms and policy frameworks. However, there were also evaluations which were less positive, finding initiatives implemented in an incoherent and piecemeal way. Gender policy was subject to 'evaporation' in part attributed to the lack of appropriate monitoring and evaluation methods (Moser 2005) which results in a lack of responsibility and accountability (Mukhopadhyay 2016).

Much effort post-Beijing has gone into making a case for the inclusion of gender issues and approaches into development processes, by arguing that it benefits the organisation and helps meet official policy objectives. Gender specialists have developed frameworks and tools, including gender assessment methods and gender budgets, as well as providing training courses on how to mainstream gender in a sector. However, to broaden mainstreaming from technical efficiency towards a more transformative agenda has been less readily accepted. A focus on technical solutions may be driven by international development agencies needing to achieve measurable results which require quantitative approaches (Parpart 2014). Changes and challenges to gender relations directly confront power, authority and equality—in this case between men and women, but gender is not the only context where these notions exist. Wherever power is confronted, it is resisted. Hence there is little enthusiasm for discussions of gendered structural inequality and the part gender plays in allocating power and resources if this is likely to draw parallels with wider arenas such as the global economy. These concepts are part of political agendas related to rights which are at odds with development conceived of, and espoused by many international agencies, as a planned process towards economic growth (which is also a political agenda). Mukhopadhyay (2016) makes the point that there is a lack of accountability in the international development agencies which allows for the transformative agenda to be sidelined.[6] While others consider that the concept of gender has become ahistorical, diluted, depoliticised and decontextualised (Cornwall et al. 2004; Mukhopadhyay 2016). Nevertheless, the use of the phrases 'gender mainstreaming', gender equality and women's empowerment is considered well embedded in the lexicon of development agencies (Parpart 2014).

An illustration of this embeddedness in the development discourse comes with the Millennium Development Goals, a United Nations initiative established in 2000 as part of the international community's response to delivering sustainable development. There were 8 goals with 21 targets to be reached by 2015. A significance of these goals is that they help frame the international political agenda for development assistance, in other words, where donor countries direct their funding. Goal 3 was to promote gender equality and women's empowerment. The goal and the relation with the other goals were not without their critics. While the overall goal has a positive tone, the target and its indicators are rather narrow which is not surprising since the process of development of the goals did not involve women's organisations but appears to have been a top-down formulation by a small number of UN officials in New York (Cornwall and Rivas 2015). Gender equality is reduced to gender parity in education at all levels with three dimensions to be monitored related to education, employment and political representation (See Box 2.1). While meeting all three targets can contribute to gender equality and women's empowerment or provide positive examples of the goal in practice, there is no guarantee that improving women's education levels and increasing employment opportunities automatically lead to the removal of other forms of inequality (Kabeer 2005; Saith 2006). Norway, for example, has no gender gap in education and women have access to well-paid employment but still experience domestic violence (Sen and Mukherjee 2014). There was silo thinking in the construction of the goals—the issues identified cut across the goals and their solutions and outcomes are interlinked.

Box 2.1 Millennium Development Goal 3

[6] Rao and Sandler (2016) rightly point out that this is not a situation exclusive to gender issues.

Goal 3: 'Promote gender equality and women's empowerment'
Target 3A: 'Eliminate gender disparity in primary and secondary education, preferably by 2005, and in all levels of education no later than 2015'
Indicators:
3.1: Ratio of girls to boys in primary, secondary and tertiary education
3.2: Share of women in wage employment in the non-agricultural sector
3.3: Proportion of seats held by women in national parliament

This distinction between technical and political mainstreaming aligns with the notions of meeting women's practical needs and strategic interests through development planning interventions (Molyneux 1985). Practical needs relate to the gender division of labour in households which for women primarily relates to their reproductive functions and activities that keep the household running and ensure their family's daily survival. While interventions to support practical needs can reduce the drudgery of these tasks and save women's time, they do not challenge gender roles and relations embedded in power relations (Moser 1989). Strategic interests relate to how women's concerns, such as gender equality, are realised. Obtaining gender equality potentially opens many possibilities, so achieving it serves women's strategic interests, but what that entails depends on the nature of women's subordination which is rooted in their social, political and cultural contexts. This variation in women's contexts serves as a reminder that women are not a homogeneous group. Women vary across a range of social characteristics such as class, ethnicity, age, religion and civil status which means that while women may share some common interests, they will vary considerably in their specific strategic interests due to their social differentiation. It should not be forgotten that power relations also exist between women. For example, younger women in India have more limited decision-making powers than older women—not only between households but also within (Agarwal 1997). In Nigeria, women's lives are shaped by relations of seniority, status and consanguinity with other women (Ogundipe-Leslie 1994 cited in (Cornwall and Rivas 2015: 403).

The need to separate 'practical needs' and 'strategic interests' has been questioned since material resources can be required to fulfil a strategic interest (Wieringa 1994). Other researchers had disaggregated 'practical needs' into productive and reproductive activities (such as housekeeping and caring tasks) which were incorporated together with 'strategic interest' into a gender analytical framework which became known as the Harvard Matrix. The framework was adopted by many development agencies for identifying gender roles in their projects (Overholt et al. 1985).

The MDGs were replaced in 2015 by the Sustainable Development Goals (SDGs), again with a specific gender goal (SDG5) (See Box 2.2). There was a mixed reaction from feminists about the content of the goal which reflects the emergence of the third phase in the gender discourse: social inclusion. While the Agenda 2030, which the SDGs are part of, is framed in terms of human rights the SDGs do not use that conceptualisation (Esquivel and Sweetman 2016). There are also concerns that policy evaporation will occur when moving down to the national level for implementation where goals and targets can be selectively applied. Nevertheless, Goal 5 has been formulated using gender analysis so has a better understanding of the causes of gender inequality (economic, social, political and environmental) and its consequences which is a step forward from MDG3, although there is no target date set for achieving the goal. Another positive is that the goals are mutually reinforcing: gender equality plays a role in achieving the other SDGs while achieving the other SDGs contributes to achieving gender equality. For example, 'equal pay for equal work' is a target under Goal 8.[7] In other words, you need more than gender parity in education to remove gender inequality which was the key target of MDG3 (Razavi 2016). The Goals are universal—so they apply just as much in the North as in the South with Goal 17 being about partnerships, so there are opportunities for mutual learning and support between the North and the South.

Box 2.2. SDG 5: Achieve Gender Equality and Empower All Women and Girls

Targets

- End all forms of discrimination against all women and girls everywhere.

[7] SDG 8: Promote sustained, inclusive and sustainable economic growth, full and productive employment and decent work for all.

- Eliminate all forms of violence against all women and girls in the public and private spheres, including trafficking and sexual and other types of exploitation.
- Eliminate all harmful practices, such as child, early and forced marriage and female genital mutilation.
- Recognize and value unpaid care and domestic work through the provision of public services, infrastructure and social protection policies and the promotion of shared responsibility within the household and the family as nationally appropriate.
- Ensure women's full and effective participation and equal opportunities for leadership at all levels of decision-making in political, economic and public life.
- Ensure universal access to sexual and reproductive health and reproductive rights as agreed in accordance with the Programme of Action of the International Conference on Population and Development and the Beijing Platform for Action and the outcome documents of their review conferences.
- Undertake reforms to give women equal rights to economic resources, as well as access to ownership and control over land and other forms of property, financial services, inheritance and natural resources, in accordance with national laws.
- Enhance the use of enabling technology, in particular information and communications technology, to promote the empowerment of women.
- Adopt and strengthen sound policies and enforceable legislation for the promotion of gender equality and the empowerment of all women and girls at all levels.

What seems to have been considered a success was the way women acting in a range of guises (civil society organisations and academics) were able to actively engage in the preparatory processes leading up to the formulation of the goals. Indeed, their work ensured that there was a separate goal since there was some pressure to have gender equality subsumed into more general 'social inequalities' (Kabeer 2016). The SDG formulation process was much more participatory than the formulation of the MDGs. Although, some participants (such as big business) seem to have had more influence than other stakeholders, such as women's groups (Razavi 2016). This helped answer one of the ongoing debates about how to engage with development agencies in gender mainstreaming. Post-Beijing saw a flurry of gender units established in development organisation. Gender equality appears to be more acceptable when it has an 'extrinsic value', for example, that women and girls can be 'change agents', rather than having an 'intrinsic value'. Some have argued that working within, it is possible to change mindsets and advance a gender-equality gender by 'speaking the language of the organisation' (Rao and Sandler 2016). While others consider that gender mainstreaming has become technocratic, since the people designated with the responsibility for mainstreaming are often divorced from the politics of feminism and so lack the analytical understanding of gender (Daly 2005). On the other hand, an analysis of the implementation of gender mainstreaming policy and praxis in the South African UNDP office between 2002 and 2006 found that the junior staff were committed, but senior management was aloof and disinterested. As a consequence of management attitudes, there was little effective gender mainstreaming programming on the ground (Joseph 2009 cited in Parpart 2014). The response to these organisational issues has been a technocratic one: capacity building through gender training, more gender tools and better planning and coordination. Little attention is given to addressing the working, sometimes hostile, environment encountered in implementing gender mainstreaming (Parpart 2014).

There has been a tendency to equate gender mainstreaming with increasing the numbers of women in an organisation in part driven by a checklist approach to evaluation—when counting bodies becomes a measure of success. While creating opportunities to women for interesting and well-paid job is to be welcomed, increasing numbers of one group at the cost of another is not always problem-free and does not always lead to the desired outcomes. Having women as the majority in a sector also does not necessarily improve women's overall position. Parpart (2014) cites the situation in the Caribbean where there is female dominance in educational achievement, but women continue to experience problems with employment and job security. To be accepted into a male-dominated organisation can require a woman to behave like men—reinforcing patriarchal norms (Catalyst 2023). Usually, a woman is given the task of being the organisations gender focal point and the gender office staff will all be women. However, there is no conclusive evidence that women are more likely to represent women's interests than men are (Childs and Krook 2009 cited in O'Neil and Domingo 2016). Placing the burden for achieving gender equality on the shoulders of women takes away any responsibility from men, even though men still dominate economic, political and cultural institutions around the world. Often at the behest of donors, women are recruited into organisations to promote gender without any consideration of the context. For example, Mukhopadhyay reports on a situation in a donor-funded project in Yemen, where there is strict segregation of women and men, which required the establishment of a women's unit in a Ministry. The women employees were seen as 'illegitimate occupants of public office because they are women and not men' (Mukhopadhyay 2016: 82). What this example illustrates is a challenge to the core values of people rather than the norms of an organisation (de Jong 2016).

Another manifestation of the dichotomy of gender mainstreaming is the confusion about whether gender mainstreaming is a goal or a tool. Daly has also pointed out that gender mainstreaming is both a concept and a policy approach. Others have questioned what the meaning of 'the mainstream' is or 'what it entails' (de Jong 2016). Is it sufficient to ensure that gender issues are central to the existing development agenda with more gender aware practice, or should we be transforming the mainstream to remove inequalities? In 2014, the Council of Europe[8] aimed to do both by defining mainstreaming as: "the (re)organisation, improvement, development and evaluation of policy processes, so that a gender equality perspective is incorporated in all policies at all levels and at all stages, by the actors normally involved in policy-making" (The Council of Europe 2014: 13).

The emergence of rights-based approaches in which women participate in public spaces as citizens not in their social identities of mothers, wives and daughters has provided an alternative to mainstreaming (Mukhopadhyay 2016). Women's rights, which encompass economic, political, cultural, reproductive and sexual rights (including bodily autonomy and integrity), were given official international recognition as human rights during the 1993 UN World Conference on Human Rights (Sen and Mukherjee 2014). In this context CEDAW[9] (Convention on the Elimination of all forms of Discrimination Against Women) adopted in 1979 is seen as having laid the groundwork for this significant milestone

[8] The Council of Europe consists of 47 member states who have signed the European Convention on Human Rights. https://www.coe.int/en/web/about-us/who-we-are (accessed 5 October 2020).

[9] https://www.ohchr.org/EN/ProfessionalInterest/Pages/CEDAW.aspx (accessed 12 October 2020).

in women's empowerment. Signatories of these international conventions are committed to reforming constitutions and civil codes to ensure women's rights. Human rights are universal, indivisible and interdependent and apply in both the public and private spheres. Given that gender is a social construct with considerable diversity between and within countries, a rights approach allows for a contextualisation of demands for gender equality (Molyneux 2004). A rights-based approach also stresses the interdependence and indivisibility of the underlying causes of gender inequality and the solutions (Sen and Mukherjee 2014) which would avoid the silo construction of the MDGs. Underpinning the Human Rights framework are three concepts which offer the possibility to address some of the criticisms outlined above: accountability, inclusion and non-discrimination (Cornwall and Rivas 2015). These concepts can be applied at any level, in any situation, organisation or institution.

The incorporation of women into the workforce opens the opportunity for financial rewards and contributions to decision-making. Feminists expressed concerns about the nature of women's work in some industries as 'underpaid, degrading and exploitative' (Pearson 2004). Nevertheless, international agencies reacted positively to women's integration into the formal economy, at least at the policy level, illustrated by the World Bank enthusiastically proclaiming that 'gender equality is smart economics' (World Bank 2007). Although it appears that the slogan did not have significant impact on the understanding of most of the Bank's economists about the social realities of many people's lives (Elson 2012). There has been continued resistance to opening the 'black box' of the household since it is assumed that a household is a unified entity pooling resources and whose preferences can be expressed in terms of a single utility function. This is countered by Sen who considers that households are social arrangements regarding who does what, who gets to consume what and who takes what decisions—which are arrived at through a combination of cooperation and conflict (Sen 1999: 192). Intra-household negotiations do not usually take place between equals. In most societies, men have more power than women to make decisions about, and exercise control over, not only their own bodies, lives and resources, but also that of other family members.

It is also contested as to whether women earning income 'empowers' them. When women enter the formal workforce, their household responsibilities do not automatically decrease. Pearson also points out that another aspect of women's traditional responsibility for the well-being of household members has become more stressful through monetisation and commoditisation of a range of goods and services such as school fees and clinic charges (Pearson 2004).

The Beijing Conference also promoted the concept of women's empowerment. This has proved as problematic as 'gender mainstreaming' in finding a consensus about its definition. Cornwall and Rivas (2015) identify three strands. Firstly, empowerment is more than giving women an improved skills set, women's empowerment is about changing power relations. Secondly, power relations are situated and contextual, they exist at all levels in society not just within the household. Thirdly, empowerment is a process not an endpoint. Skutsch et al. (1999) posit that a lack of clarity in defining empowerment leads to projects that incorporate women's empowerment as a goal will struggle to reach their objectives. Practitioners are probably consciously or unconsciously interpreting empowerment to mean 'having the power to make decisions and solve problems' and/or 'having the power to organise with a common purpose or common understanding to achieve collective goals'. The latter definition links to a feminist notion of women's collective action (or 'power with') which manifested itself in the WID development discourse. However, collective action requires having the time and freedom of movement outside the home to participate which would prove problematic for some working women and women from some ethnic and religious groups. Feminists also argue that for women to be empowered, they need 'power within', that is, self-confidence, self-awareness and assertiveness which enable them to analyse their experiences, to see how power operates in their lives, and so gain the confidence to act to change their situation, to challenge those people or organisations and institutions, such as the state, who hold 'power over' them (Parpart et al. 2002). Having 'power within' creates a sense of agency, which has been defined as having the capacity to define one's own life-choices and to pursue one's own goals, even when faced with opposition from others (Kabeer 1999). In line with this view, Moser considered that empowerment is the ability 'to determine choices in life and to influence the direction of change, which required the capacity to gain control over crucial material and non-material resources' (Moser 1993: 74–75). Having self-confidence and an awareness of one's possibilities encompasses what Friedmann refers to as psychological power (Friedmann 1992 cited in Standal 2008: 23). However, Friedmann considers that power is multidimensional and that for someone to take control over their own development, requires access to different types of power—not only psychological but also social and political power. Social power enables an individual or group to increase and enhance capacity to decide upon and meet objectives which is contingent on the possession of a range of assets (such as information, knowledge, skills and finance as well as participation in social organisations). Political power is through participation in decision-making processes in informal and formal institutions, voicing opinions and taking collective action which provides the 'power to' bring about change (Parpart et al. 2002). Friedmann's three dimensions of power are considered to be contingent upon each other and are mutually reinforcing.

Having access to and control over resources are also an important contribution to determining one's personal development. Resources consist of a range of forms including physical, human, intellectual and financial resources (Sen 1997 cited in Cornwall 2016: 344). However, Batliwala (1993 cited in Cornwall 2016: 344) adds that one also needs to have control over ideology, that is, beliefs, norms, values and attitudes, since these confer power and authority to certain individuals who can control access to and enjoyment of the benefits of resources. For example, in some rural societies, a woman will have the right of access to her husband's land to grow crops or graze animals, but she will have no rights to determine whether that land is sold—that is her husband's prerogative.

There is concern, within the development discourse, that women's empowerment has been reduced to economic empowerment. Women's economic empowerment is seen as a duality. It is good for the economy, for example, it was one of the main messages of the 2012 World Development Report which took Gender and Development as its theme (World Bank 2011), but it is also the route to the removal of all other dimensions of inequality. 'Empowerment becomes an individual resource to be maximised for efficiency' (Cornwall and Rivas 2015: 406). However, empowerment is a multidimensional process that follows several pathways (personal, economic, political) and engages across a range of scales: individual (mind and body), household (family) and society (community, organisations and government) (Goldman and Little 2015). Nevertheless, development projects tend to focus on only one sphere, often at the local level, and scale of engagement using measurable outputs, for example, the number of girls in school, women as entrepreneurs or women elected to government assemblies.

Figure 2.1 shows around the start of the twenty-first century a third phase in how gender appears within the development discourse: social inclusion. Feminist and social scientist researchers had been arguing for some time that gender is not a binary condition but that individuals are differentiated by a range of social and economic characteristics which need to be taken into account. Indeed, life is more complicated than reducing social relations to one of male dominance and female vulnerability (Cornwall and Rivas 2015). In the 1990s, black feminist researchers in the US argued that there is an interconnectedness and interdependence between race and other social categories such as gender, ethnicity, socioeconomic status, sexual orientation and other social identities or positions which shape social experiences (Valentine 2007) or are affected by economic or political projects (Yuval-Davis 2016). Social categories can mutually strengthen or weaken each other (Winker and Degele 2011). To describe this situation related to social categories the legal scholar, Kimberlé Crenshaw, is credited with introducing the concept of intersectionality (Crenshaw et al. 1995 cited in Valentine 2007: 12). The concept has been used to highlight how women in different positions (for example, immigrant women) and identities (for example, LGTBIQ) experience and are affected by the same phenomena and policies in different ways (Paulson 2016). Intersectionality is used as a lens to focus on different groups without prioritising one facet or category of social difference (Yuval-Davis 2016). There are researchers who use intersectionality as originally intended, that is, to analyse the specific situation of black and ethnic minority women (Yuval-Davis 2011 cited in Paulson 2016: 399). Other researchers have broadened its use as a tool to analyse social stratification revealing positions that are marginal in some ways but dominant in others, for example migrant men or white women, with some arguing that it helps reveal the role of 'white men' who tend to be hidden. Indeed, men appear to have literally been left out of the picture altogether, as Cornwall and Rivas (2015) wrily remark that images of empowered women—running their businesses, speaking in public, sending their daughters to school—in promotional material tend to be completely devoid of any men. The invisibility of men has led to the establishment of a research area which applies gender analysis to men, examining notions of masculinity within and between societies, reminding us that men are not a homogeneous group (see, for example, Hearn and Morgan 2014; Lugones 2016; Pulé and Hultman 2021). Not all men are dominant oppressors, so we need to examine positive ways of incorporating men into the development process to support gender equality (see, for example, Cornwall and White 2000). Latin American scholars have raised awareness about how colonialism installed institutions in their colonies which were rooted in a patriarchal system that have shaped gender relations in the mould of the coloniser (see, for example, Lugones 2016). The introduction of colonialism into the gender and development discourse has contributed to opening up the debate with several lines of criticism emerging. Colonisation has shaped all aspects of human life in a way which reflects the worldview of wealthy, white, European Christian men (Paulson 2016) which, by extension, devalues indigenous identity, knowledge and different interpretations of the place of human beings in the natural world and the cosmos (Middleton 2010).[10] Evidence of the impact such a system has had on women's status comes from research about Yorùbá society and culture, which extends across the contemporary states of Nigerian, Benin and Togo. Before colonialism, there was no gender system in Yorùbá society, however after colonialisation women became defined in their relation to men and were unable to have power, own land or participate in leadership roles in society (Oyěwùmí 1997). The development process is also dominated by this colonial worldview stifling other visions of development. Although these alternative visions find their expression in social movements (see, for example, Escobar 1998; Tamma and Duile 2020). GAD has been criticised for presenting gender as a universal homogenous experience of female subordination instead of one which is contextually nuanced. Collective identity in some cultures is stronger than individual identity (Hirtenfelder 2015) and in others gender is not constructed around anatomical distinctions (Walsh 2016).

For development practice, the concept of intersectionality points to the need for data collection and analysis to reflect differentiation. However, there has been quite a debate among researchers about a lack of clear

[10] Linked to the heterogeneity of women, further challenges emerge from Latin American where in the 1990s Marxist analysis introduced the term coloniality which describes the notion of a hierarchical social classification of people in terms of their race as imposed by European colonialism in which indigenous people were made to feel inferior and used as a means of social domination (Quijano 2000 cited in Lugones 2016). However, this analysis has been criticised for the way in which it conceptualises gender and race as separate from each other and so misses a rather Eurocentric perspective which is overly reliant on a particular biological construction that 'presupposes sexual dimorphism, heterosexuality, and patriarchal distribution of power' (Lugones 2016: 18).

definition of the concept which then creates uncertainty about the categories to use and where to draw the boundaries (Davis 2008). Perhaps, for an organisation, intersectionality can act as an enabler to engage critically with assumptions about the population or community which a policy or project aims to serve and determine which categories to consider when collecting and analysing data—to provide at least internal consistency. For the individual researcher and practitioner it can be seen as a mechanism which 'force us to examine the extent to which we harbour assumptions, myths, stereotypes and limiting beliefs that prevent us from treating everyone with dignity and respect' (Cornwall and Rivas 2015: 410) and 'how the identity of the researcher/s can influence participants because of their own identities and historical experiences' (Hirtenfelder 2015: 35).

International organisations have used intersectionality as a tool for addressing social inclusion which had become part of the twenty-first-century development agenda linked to human rights that are set out as part of the founding principles of the United Nations.[11] These rights are universal fundamental entitlements that a person can expect and exercise, such as the right to a nationality and to vote, in the knowledge that they are protected by the law. When a person or group feels that they are denied these rights then they may take action to demand justice (Nussbaum 2000). For example, if people do not have access to the resources and recognition that would enable them to participate fully in the society where they live, they may demand 'social justice' to create a fairer system (Kabeer 2012b). Achieving social justice requires the existence of institutions that those seeking redress have access to and that are operated by fair and impartial decision-makers (Sovacool and Dworkin 2015). The policy response to demands for social justice is to promote inclusion of those groups who can be categorised as socially excluded. However, such responses have been criticised for a rather simplistic understanding of the situation of those calling for social justice, for example, ignoring or underestimating power relations which have shaped the differentiation, cannot be expected to easily reduce differentiation or inequality in society (van Berkel et al. 2002; Hospes and Clancy 2011). Indeed, the concept of social inclusion has be criticised for being misleading, contested and elusive (Hospes and Clancy 2011). Those who are targeted for inclusion may have their own reasons for remaining excluded. The terms of

[11] UN Declaration of Human Rights. https://www.un.org/en/about-us/universal-declaration-of-human-rights (accessed 21 January 2022).

inclusion can be problematic, for example, they may reduce the impacts of exclusion but not address the causes and institutional mechanisms of social exclusion (Kabeer 2000). Kabeer points out that 'the problem for women was not so much that they had been excluded from development, but that they had been incorporated into it on adverse terms' (Kabeer 2000: 87).

Instrumental in the demand for women's rights in the Global North was the feminist movement of the 1960s which, reflecting the diverse composition of the movement, had different strands of interest including access to education, improved work opportunities and employment conditions and sexual and reproductive autonomy as well as challenging misogyny and patriarchy (Howard-Hassmann 2011).[12] During the 1980s, women from black and indigenous communities began to challenge the feminist movement for its focus on individual rights neglecting differences between women across the globe and raising the issue that different social and cultural contexts might have different issues. The context also mediates women's access to resources. While the legal system formally acts as a guarantor of rights, many women are subject to informal social power structures and resolution of disputes particularly between a woman and a man. There were women who argued that since their injustices were rooted in gender norms, they needed 'patriarchy'—a concept which emerged in the 1990s. Not surprisingly given these differences in interpretation of what constitutes 'women's rights' there was also disagreement about what constitutes 'gender justice'. Some questioned whether the legal system could deliver gender justice since the patriarchal mindset and social relations within households and communities are also in operation in other institutions such as the legal system which is dominated by men. Indeed, a universally agreed definition of gender justice is elusive and rooted in different discourses (Goetz 2007).

There are understandably frustrations arising about the slow progress with the outcomes of gender mainstreaming linked to the significantly powerful institutional resistance, including within the UN system, often by conservative religious groups (Molyneux 2004). The UN system operates by trying to find a consensus between competing actors from

[12] Although of course women, with support from men, had been demanding their rights far earlier than the 1960s as can be seen, for example, in Mary Wollstonecraft's landmark publication in 1792: A Vindication of the Rights of Woman.

development agencies, national governments and civil society organisations which can lead to a watering down of aspirations. Unfortunately, the experiences in Sweden, a country to which many outsiders look to for leading the way on gender and social justice, do not offer hope of progress beyond a snail's pace. Sainsbury and Bergqvist (2009) describe how it took more than twelve years to mainstream gender into all the ministries and the Prime Minister's Office. This situation is despite a commitment to mainstreaming by the ruling party. However, a look at history shows how long the struggle for women's suffrage took. It was only in 1971 that women in Switzerland achieved the right to vote in federal elections, and later for some lower governmental levels,[13] which is nearly 100 years after New Zealand granted full universal suffrage to women.[14] In this chapter, the focus has been on development as it takes place within the UN system but, it should be pointed out, that the fight for gender equality and women's empowerment takes place in other arenas as well. The internet is opening up a global space for alliance building in which women are able to join across international boundaries which a few decades earlier would have been financially prohibitive (Moghadam 2000; D'Enbeau 2011).

2.5 The Gender and Energy Nexus in the Development Discourse

2.5.1 The Early Years

While Ester Boserup can be credited for bringing women as an actor into the development discourse, that distinction for bringing global attention to women in the specific context of energy could go to Eric Eckholm with his seminal report in 1975 for the World Resources Institute in which he used the phrase 'the other energy crisis' (Eckholm 1975). The term 'crisis' was used to draw a parallel to the oil crisis referred to at the start of this chapter: the majority of the world's population faced problems in accessing their main energy source: fuelwood. The primary response from the International Development Community to this crisis in the South was support for increasing fuelwood supply through a range of

[13] https://www.parlament.ch/en/%C3%BCber-das-parlament/political-women/conquest-of-equal-rights/women-suffrage (accessed 12 October 2020).

[14] https://nzhistory.govt.nz/politics/womens-suffrage/world-suffrage-timeline (accessed 12 October 2020).

options including farm-based woodlot projects. A driver of this approach was the environmental concerns that fuelwood use was causing deforestation (Leach and Mearns 1988), although this generalisation has been challenged with evidence pointing to commercial timber being the major culprit (Arnold et al. 2003; Masera et al. 2015). The intervention logic for promoting woodlots was based on identifying women as the main fuelwood collectors and users within their households. Such an intervention was advocated as a direct way to help women save time. Indeed, Irene Tinker, a gender specialist, had argued that women's time was the real 'rural energy crisis'—their long working day based on their physical labour to complete many household tasks involving drudgery, such as grain processing, left them with no time to participate in projects (Tinker 1987). The solution was to provide improved mechanical energy devices to reduce the time and effort used in these tasks which are important for the household's welfare (CRGGE 2006).

During the 1980s evidence was pointing to the gap between fuelwood supply and demand not decreasing, in other words these policies were not working (Arnold et al. 2003). In the context of woodlots, evidence from research began to emerge which showed that there had been a failure to take into account the social relations of gender. For example, women usually did not own the land, they did not have the right to plant or cut trees, or, once the trees had been cut, they had no control the use of the resource—men owned and controlled the resources (Leach and Mearns 1988). In other words, gender relations directly impacted on energy outcomes.

One of the outcomes of the attention to the 'fuelwood crisis' was a focus on rural energy in the South which highlighted the role of women in relation to household energy. However, this was a narrow interpretation of women's relationship with energy which was taken as synonymous with cooking—men's energy needs were not specifically identified. The response to energy for cooking was primarily a technical discussion about the potential for improved energy efficiency in both fuels and appliances such as stoves. This discussion was usually conducted by male engineers from the North, while the primary uses of the technology, women in the South, were excluded from the design process.[15] However,

[15] A similar story exists in the development of electrical household appliances in Europe and the US—men designed what they thought women needed—although women, particularly in the US, were used to promote appliances (Matly 2005).

organisations with field experience in the South, for example, the Foundation for Woodstove Development and the Household Energy Network (HEDON) pointed to their experience in which there was a wider uptake of stoves and a more sustained use if women were involved in the design process itself (McDade and Clancy 2003). Attention to wider issues began to appear, for example, Cecelski, while working with the International Labour Organization (ILO) during the 1980s, pointed out that women also used traditional fuels for productive activities (Cecelski 1987).

Nevertheless, energy in the context of development received little attention beyond 'traditional fuels' of wood, agricultural residues and animal dung with women portrayed as managers and users of natural resources. The metabolic energy of physical labour in household survival activities was ignored. This was despite the existence of evidence which showed (using time as a proxy indicator for metabolic energy) that women spent more hours than men on meeting basic needs which has a negative impact on women contributing to household income and food production (Cecelski 1995). There was hope that the Beijing Conference would address these issues. In the lead up to the conference, in an attempt to broaden the concept of gender and energy, Cecelski published a paper in the influential journal 'Energy Policy' (Cecelski 1995). She argued that the development agenda which focused on the fuelwood crisis of stoves, time-saving, wood-lots and biomass fuels, needed to expand to one that encompassed a broader range of issues including pricing, transport and modern energy forms, such as electricity and LPG. Policymakers did not appear to be picking up these messages with women/gender and energy not featuring in the Beijing Conference.

Researchers decided that, if progress was to be made with getting gender and energy onto the international agenda, a more strategic approach was required by presenting evidence in a focused, practical and coordinated way. So, in 1995, a small group of women from the Global North who had been working on the social dimensions of energy came together to establish an international network on women and energy: ENERGIA.[16] The intention has been to bring together researchers, development experts, energy experts and gender specialists from both the South and the North to exchange experiences, generate data and share

[16] The name was later revised from 'women and energy' to 'gender and energy'. See www.energia.org (accessed 22 April 2022).

analytic approaches. The network aimed to bridge the gap and stimulate dialogue between two different groups of actors who can sometimes appear to be running on different tracks: firstly, policymakers and those working in development practice at the grassroots and, secondly, energy professionals and gender experts. Network members began generating the evidence about gender issues related to energy use, much of which has been drawn on in writing this book. Policymakers were targeted with information through a newsletter (ENERGIA News) which covered a range of issues beyond stoves. There were also training manuals developed to explain how to mainstream gender into energy policy, projects and practice.[17] A sign of recognition that the message was getting home was the publication of training manuals by international development agencies.[18]

There have been differences in interests between academic researchers and development practice. The latter group has focused on two main streams: cooking and productive uses of energy. Cooking has focused on both the technology and the fuels with efforts to promote LPG (see, for example, Smith 2002; Cecelski and Matinga 2014) and electricity (see for example ESMAP 2020). The approach at promoting the transition to clean cooking was to treat women as passive beneficiaries with little effort to engage them in the design process or understand the way gender relations operate at the household level. With support from the World Health Organisation (WHO), a substantial body of scientific evidence was built up in relation to the health implications for women of working in smoky kitchens (see for example, Smith 1999). Wickramasinghe (2003) pointed out that there were other neglected health issues related to the entire fuelwood cycle from production, through collection and combustion, to post-combustion issues. These issues continue to be neglected (Matinga and Clancy 2020). Indeed, development practice measured success in terms of stoves disseminated (which as we will see in Chapter 3 is not quite the same as stoves used) rather than the beneficial health impacts (CRGGE 2006).

Much of the early work focused completely on women who were treated as an undifferentiated group. Men and their energy requirements were virtually excluded from the picture. What helped to bring men into

[17] See, for example ENERGIA (2007).

[18] See, for example, UNDP (2004), Angelou et al. (2019).

focus was attention switching to looking at the entire stoves chain where it could be seen how men dominate from decision-making bodies, such as energy planning commissions, in technical and economic research, occupy higher-status and more remunerative commercial activities, such as charcoal production and distribution, and as artisans, for example, men were the main stoves producers (Clancy and Feenstra 2006; CRGGE 2006). Looking back at the previous section with the comment about missing men, one could say that it was not so much that men were missing from the picture, they were the picture—the male was the norm.

A more comprehensive view of household energy emerged towards the end of the 1990s. Household energy involves more than cooking and stoves but encompasses all the activities that take place within a household and the linkages to a much wider system of energy supply and demand as well as to other sectors (see for example Clancy 1998, 2002; Klingshirn 2000). The household was also recognised as a centre of production which could involve the family's cookstove, particularly for women, meaning that activity needed to be considered when designing stoves (CRGGE 2006). It was not only stoves that needed attention but also cooking practices and food preparation, as well as kitchen and house design which required an understanding of cultural practices (Clancy 2002). Generally, women have responsibility for energy provision related to their spheres of influence, particularly for activities centred on the kitchen. However, fuel is only one part of a more complex interconnected system related to the kitchen which also involves the physical structure and location of the kitchen, the devices and equipment, cooking practices and food (Dutta 1997). Men and women play very distinct roles related to the gendered division of labour based on cultural practices which also influence decision-making around the components of the system. For example, men assume responsibility where fuel has to be collected over long distances, where fuel is purchased, or social norms restrict women leaving their homes. Men also make decisions about the type of stove to buy, or they prioritise other equipment, such as TVs and radios, over improved stoves (Makan 1995) or reject a particular type of stove which they do not consider meets cultural practices (Wilson and Green 2000).

Social scientists pointed out that development practice was taking a narrow view of what constitutes a household tending to equate it with the nuclear family (Clancy 2002). However, the family unit forming a household varies from culture to culture. It can consist of several generations of people connected by marriage or be wider including people who have no kinship relationship with the family, such as, servants, workers and paying lodgers. Household composition is also fluid with adult male members migrating to find work, which can impact on household income and on gender roles. For example, in Nepal, rural male migration is part of a household strategy to earn income and women are left to manage the farm, taking decisions about crops and livestock that in the past have been by men. Women have also become more mobile outside of the house and village which is increasingly accepted as a social norm (MSSRF and CRT Nepal 2019).

To summarise, the first attempts to address gender issues related to energy were a narrow focus on household energy defined as cooking with wood stoves identified as the main culprit of deforestation which became the focus of attention. Solutions took a largely technocratic approach, without involving women as the end-users in stove design or whether cooking was women's main problem in search of a solution. Men's role in the household, particularly in relation to decision-making, was ignored, as were other aspects of cultural practices. Urban areas received less attention than rural areas even though urban women's inequalities are not dissimilar to those of rural women. Compared to men, women receive lower remuneration in the labour market, own lower levels of productive assets and resources as well as suffering exclusion through social stigma and discrimination (Amis 2002).

2.5.2 The New Millennium

Development practice in the new millennium was influenced by the MDGs which, as pointed out earlier, had a gender goal but no energy goal. However, WSSD did acknowledge that access to energy is needed to help achieve the MDGs. Another important recognition emerging from WSSD was that gender and energy was more than women and stoves. Another change came in 2005, with a review of the Beijing Platform for Action making explicit reference to investment in energy infrastructure having the potential to reduce women's and girls' time spent on fuelwood collection and transport. A new indicator was added to Goal 3 on gender equality which is to measure the hours per day women and men spend fetching water and collecting fuel. The formulation of this indicator is significant in that it does not see these reproductive tasks as solely women's responsibility.

Development practice drew on the sustainable livelihoods framework which had originally been advocated by Chambers and Conway in the early 1990s (Chambers and Conway 1992). This framework had three elements. Economic activities were one element, which fitted with the dominant paradigm of development, the other two elements had a stronger social dimension in contributing to human capabilities (such as education, health and skills) and access to tangible assets (such as finance) and intangible assets (such as kinship, community and political networks). The household is the centre of the analysis examining both inter- and intra-household relationships. Several studies were undertaken to understand the role that energy played within the livelihoods framework. The focus tended to be on the supply side in which the conclusion was that a lack of sufficient energy sources (both fuel and labour) contributes to a 'vicious circle' of poverty, which could be reverse into a 'virtuous circle' by the provision of more energy.

Initially intra-household relationships did not seem to receive much attention. However, a study funded by the UK's Department for International Development (DfID) put gender at the centre of the energy-poverty debate (Clancy et al. 2003). The authors argued that energy poverty should be considered a core dimension of poverty and that gender was very much a part of this nexus since the majority of the people living in poverty were women whose needs, constraints and strategies for leaving poverty were different to men's. They explained that the sustainable livelihoods framework helps unpack the relationship between gender and energy services, illustrating their arguments by showing, using the Harvard matrix, how different energy forms can contribute to meeting women's different needs[19] (see Table 2.1). They challenged the unitary model of the household which assumes equality in decision-making and no differentiation in needs, priorities and constraints for individual household members. Instead, they argued to have a more comprehensive understanding about energy services in which intra-household relations needed to be part of the analysis, by asking the questions about who makes decisions about energy within the household and who benefits from that decision. The recommendations called for more women's

[19] Although, the framework was criticised for a lack of attention to small and micro-enterprises which form the basis of many households income-generating activities—a situation attributed to the framework's focus on agriculture and natural resources—which in turn makes it rural-biased (IDS 2003).

involvement in the energy sector at different levels and a focus on energy services, so a demand-side perspective to balance the dominant supply side, to give a better indication of whether energy availability was having any impacts and the nature of those impacts. There was also a call for more research to provide micro-level evidence to test the assumptions about the benefits of energy access that form the basis of policymaking, for example, women will use electric light at night to study (Skutsch n.d.). Cecelski (2004) proposed four hypotheses which could be used to provide the evidence, which tested both gender and energy as key variables in achieving policy objectives related to health, productive and reproductive activities, energy policy formulation and gender equality. These hypotheses summarise much of the research on gender and energy in the context of the Global South.

Several arguments were advanced to explain the lack of gender mainstreaming in energy projects. One argument was the reliance on the unitary model of the household as the basis for data collection for energy planning (Skutsch n.d.). If data collection relied on household surveys, these would usually be directed at the head of the household which, if there were adults of mixed sex, would be identified as the senior man. The implicit assumption is that the man responding to survey questions knows all the activities of household members and their energy needs. The energy sector did not have the benefit of local-level 'extension' staff who worked at the community level that other sectors, such as health and agriculture, had who could provide information about how households work. Utilities considered community engagement as time-consuming and not worthwhile (James 1998).

Skutsch (1998) argued that a significant cause of failure to mainstream gender into energy projects was the lack of explicit motivation for including gender as a variable in planning which in turn led to a lack of clarity in objectives, indicators, approaches, techniques and methods. As a result, there was muddled analysis leading to projects that did not match the commitments of planners to gender equality and the expectations of intended beneficiaries to improve their lives. Several explanations were offered for this phenomenon known as 'policy evaporation' (Derbyshire 2002). In some cases, this was due to staff in both donor organisations in the Global North and the recipient organisations in the Global South not being committed to 'gender equality' in a development context nor did they see the relevance of 'gender' to their work (Christian Michelsen Institute 1999). Indeed, there was some resentment towards what was

Table 2.1 Possibilities for improving the position of women through energy (Clancy et al. 2003)

Energy form	*Women's needs*		
	Practical	*Productive*	*Strategic*
Electricity	– Pumping water: reducing need to haul and carry – Mills for grinding – Lighting improves working conditions at home	– Increase possibility of activities during evening hours – Provide refrigeration for food production and sale – Power for specialised enterprises such as hairdressing and internet cafes	– Make streets safer: allowing participation in other activities (e.g. evening classes and women's group meetings) – Opens horizons through radio, TV and internet
Improved biomass (supply and conversion technology)	– Improved health through better stoves – Less time and effort in gathering and carrying firewood	– More time for productive activities – Lower cost of process heat for income generating activities	– Control of natural forests in community forestry management frameworks
Mechanical	– Milling and grinding – Transport and portering of water and crops	– Increases variety of enterprises	– Transport: allowing access to commercial and social/ political opportunities

seen as imposition of donor ideology. Another explanation was that staff lacked the commitment for mainstreaming gender into their work because they did not have the necessary skills and understanding about how to do it and the benefits it brings to achieving objectives (Derbyshire 2002). A lack of understanding of gender and intersectionality could be seen in project reports using 'women' and 'gender' interchangeably or using the undifferentiated terms such as 'people', 'community' or 'consumers' (Cecelski 2004). This lack of understanding was despite, post-Beijing, tools available for gender mainstreaming into projects (such as the Harvard matrix and the gender analysis matrix) which were already in use in the forestry, agriculture, water and health sectors and yet these tools were not used by energy planners (Clancy et al. 2003). It was argued that this lack of enthusiasm for gender tools was because the existing tools were too general and did not focus on the specific energy dimensions of poverty (Skutsch 2005). This realisation resulted in a lot of effort by NGOs from development practice, researchers and development agencies to produce training manuals for mainstreaming gender into energy policy and projects (see for example: Skutsch et al. 2004; Sharma et al. 2005; Havet et al. 2007; Clancy et al. 2009; Cecelski and Dutta 2011).

A lack of clarity about gender objectives or goals was considered a significant factor in energy project failures with different participants in the project having different understandings about the goals of the project (Skutsch 1998). An analysis of energy projects found that three types of energy goals could be identified: (i) improving women's welfare; (ii) improving women's productivity; and (iii) empowerment (Skutsch n.d.). These three goals line-up with the three components of the Harvard matrix shown in Table 2.1. The welfare goal addresses practical issues, the productivity goal addresses production and the empowerment goal relates to women's strategic interests. There can be overlap in goals, for example, time saved in care work (welfare/practical) can be used for setting up a small business (productivity/production) or attending a women's group meeting to campaign for rights (strategic/empowerment). The concept of project goals can help in clarifying what the objective of a project is, for example, if it really contributes to women's empowerment. Indeed, Skutsch (n.d.) considered that donors were reporting, unchallenged, that their project had contributed to women's empowerment when in fact the project was about women's welfare. Indeed, there was a discussion within the gender and energy community as to whether energy projects could actually empower women. The conclusion was that there was probably no piece of hardware that could directly empower women, but it was the process of acquisition which was the enabler (see, for example, Khamati Njenga and Clancy 2002; Skutsch 2005; CRGGE 2006). Cecelski came to a similar conclusion when looking for a correlation between energy use and women's empowerment. She compared per capita energy use with the UNDP's Gender Empowerment Measure (GEM), which measures, for more than 50 countries, gender inequality based on estimates of women's economic income relative to men's economic income and their access to professional and parliamentary positions. She found that there was only a very weak association between a small increase in energy availability and an increase in gender empowerment, although for small increases in energy access in low-income countries there were significant increases in other development indices such as life expectancy, literacy and income (Cecelski 2005).

Skutsch (2005) proposed an alternative approach to formulating and implementing energy projects with a gender dimension which she called 'the project efficiency' approach. In this case, an explanation for project failure is because planners do not understand what people's needs and priorities are, so involving women and men in identifying and prioritising their needs would lead to increased project success. This approach was also seen as a way of overcoming resistance by male project staff, referred to above, for gender mainstreaming into their work. Identifying needs had to be done by not relying on household surveys but by having meetings with community members which would give women a voice. The decision to hold mixed or single sex meetings would depend on local culture as to whether norms would allow women to speak in public.

The next step was to mainstream gender into the whole project cycle by asking gender-related questions at each stage (Dutta 2003b; Karlsson and Clancy 2004). The type of data generated by these questions was also significant since aggregated data at the household level missed inequalities within households in respect of needs, priorities and capacity to act (IDS 2003). The use of qualitative research methods was promoted for providing insights into women, men and children's everyday practices which would support energy project planning and implementation. For example, time allocation studies by anthropologists of farming practices in eight ethnically different districts in Nepal showed the dangers of assumption about the homogeneity of women and men's roles in rural areas with the communities studied having different attitudes to female entrepreneurship related to cultural norms and values (Acharya and Bennett 1981). Failure to understand the gender roles and relations in a particular context led to women's situation worsening. For example, projects to improve fuelwood supply by encouraging women to plant trees on land they farmed, failed to understand in Kenya that this action would under customary law grant ownership of the land, while in South Asia women lost their customary rights to gather fuelwood on communal land when the land was appropriated for community forestry projects in which men controlled the trees and their products (Agarwal 1986).

Other criticisms of energy projects began to emerge such as an over-reliance on technocratic approaches to introducing energy technologies, in which women were seen merely as passive beneficiaries of energy projects. This rather narrow view resulted in the lack of participation by women in the design process of the very technology they used and could be considered experts. Women's exclusion in design processes, as well as minimal membership of planning and decision-making bodies, were identified as causal factors in the low take-up of hardware such as improved stoves (Cecelski 2004). Evidence pointed to the benefits for women participating in projects, such as gaining technical and administrative skills which could be used to find work after the project finished (CRGGE 2006).

Initiatives were started to encourage and promote women's involvement in delivering energy access, for example, working in energy utilities, or through owning and running enterprises which sold energy technologies or new and energy-efficient equipment the remark earlier that men dominated household stove production, it was not uncommon to find women producing stoves, for example, in India (Bhogle 2003) and in Uganda (Njenga 2001). In the 1990s, the Vietnam Women's Union were marketing Solar Home Systems. Their success was attributed to the 'saleswomen' of the Union, as housewives themselves, knowing the energy needs of the female clients best (Everts and Schulte 1997). However, most of the early efforts of women as energy entrepreneurs appear to have been as part of donor-funded projects. A review of the literature was not able to identify any studies that discussed experiences of women as energy entrepreneurs established and running under free market conditions (Dutta 2003a). This result should not be interpreted to mean that there were no energy enterprises owned and operated by women but rather, as is discussed in Chapter 4, it is possible that since women's enterprises tend to be in the informal sector, often operating from their homes, they are missing from data sets.

Most of the research had focused on rural areas with only limited attention paid to urban areas taking mainly a sectoral focus with a clear neglect of gender particularly in relation to energy and livelihoods (Clancy 2006). The available evidence suggested urban and rural households were not dissimilar in terms of energy provision for the house: primarily it was the women of the household's responsibility. Urban women like their rural sisters had, relative to men, limited capabilities, lacked productive assets and resources and received low rewards in the labour market facing exclusion through social stigma and discrimination. A three-country study (Nigeria, the Philippines and Brazil) did examine energy use in an urban context with attention to the gender dimensions (Clancy et al. 2007). The researchers found that while energy availability might be better than in rural areas, poor urban households still had access challenges, not all cost-related. Wood fuels, particularly charcoal, were found to be an important part of the fuel mix, including for high-income households. There were also changes in household practices because of urban lifestyles. Paying for cooked food to eat outside the home, often at informal road-side restaurants, was not uncommon for all income groups. The causes of this phenomenon varied. A significant number of men living in urban areas have no partner and have never learned to cook. In the Philippines a small number of households reported switching to buying cooked food in response to energy price increases. In Nigeria, women were working long hours outside the household and did not have time to cook. The availability of electricity in urban areas of the Philippines and Brazil enabled a proliferation of services industries and job opportunities for women, such as clothes washing, ironing, hair-dressing and dress-making. However, these occupations are related to housekeeping skills which tend to be low-skilled with poor remuneration (Clancy et al. 2006).

2.6 The North Takes an Interest

The realisation that energy has a gender nexus took time to appear in the discourses of the Global North. Around the start of the second millennia, a small number of publications appeared which began to make the case that there was a gender-energy nexus in the North (see, for example, Clancy and Roehr 2003). Men and women play different roles in Northern societies, although at first glance these may not be so clearly defined as in the South. However, research shows that women, while remaining part of the paid workforce, still continue to carry the main responsibility for household tasks and childcare in Northern societies.

At the turn of the Millennium, the European Commission funded research to examine whether women's concerns, as well as men's, were integrated in the design implementation, monitoring and evaluation of the Commission's energy policies and programmes (Clancy et al. 2001). The research team complained that there was a lack of gender-disaggregated data which made it difficult to draw any conclusions, for example, whether women had any influence on the direction of energy research policies or within research programmes and projects. Indeed, gender was not considered an issue in energy programmes even though the Commission had started to accept that science is not always gender-neutral and that social considerations needed to be taken into account when awarding research grants. However, the study found that within the Commission there was a general lack of awareness of what gender issues and gender-oriented research are about and, as a result, a lack of capacity to incorporate these questions into the programmes.

The framing of the nexus in the Global North, at least in the European research context, has been in terms of two broad categories: gender, social inclusion and energy poverty and women working in the energy sector.

2.6.1 Gender and Household Energy

It took some time for the gender and intersectional dimensions of household energy to be recognised as issues in the Global North. Early surveys about household energy did not collect gender-disaggregated data, for example, a detailed household energy survey in New Zealand which aimed to identify the technical and behavioural factors that determine energy use (Stoecklein 1997). This lack of inclusion of gender was despite demographic evidence related to household income indicating that this was at least an issue for investigation. Available data for the US in 2000 showed that there are more families in households classified as living in poverty headed by single women than households with an adult male (24.7% compared to 4.7%) (Dalaker 2001). There were also indications of intersectional issues. For example, in the US, African-Americans, in 2000, made up approximately 12% of the population but formed 28% of the people classified as poor (Dalaker 2001).

The social dimension of energy in Europe has focused on vulnerable consumers using the concept of energy poverty rather than energy access as formulated within the SDGs. This approach reflects a perspective that considers problems related to energy access in industrialised countries are related to capacity to pay for energy rather than supply-side issues related to energy availability which are considered to be major factors in many countries of the South. Attention to the scale of the issue of energy poverty came after the groundbreaking work by Boardman in which she drew attention to the large number of households living in energy poverty[20] in the UK which she estimated, in 1991, at 7 million (36% of total) (Boardman 1998).

It came as a surprise to many in the Global North to learn that there were women, at least in Europe, who were reliant on fuelwood for cooking and heating which they had to search for and experienced the sorts of physical and psychological violence not unfamiliar to women in the Global South. This was particularly prevalent in war-torn areas or places recovering from strife, for example, Kosovo, Bosnia-Herzegovina and Chechnya (ENERGIA Support Group and CSD-NGO Women's Caucus 2001). It is also possible that these women were exposed to the same health issues as women in the Global South from household air pollution. However, this appears not to be recognised as an issue by policymakers.

There was some early evidence from research of women not forming a homogeneous group. A study in Germany comparing motivations for saving energy in the home and approaches used to meet this objective found that women of pensionable age tended to use less energy than younger women (Preisendoerfer 1999). To reduce energy consumption older women adopted behavioural changes, for example, they cooked less, whereas young women tended to opt for technological solutions. These differences in strategies for energy reduction could be a result of socialisation processes and their historical context. The older women grew up during and after Second World War which was a period of austerity which stressed frugality, reinforcing behaviour to promote saving if women from that age group find themselves living with low incomes. Younger women are more familiar with technology since it has been part of their lives, so it is a logical step to explore whether new appliances and computer technology can reduce their energy consumption without negatively impacting on their everyday practices.

There were also some household energy studies which compared women and men, with some disaggregation across social categories. In the Netherlands, a survey looking at environmental attitudes and behaviour, found that energy use per capita was greater for two-income households (defined as with a working female partner) than in households where the

[20] The UK uses the term 'fuel poverty' rather than 'energy poverty'. For consistency the term 'energy poverty' is used throughout this book unless giving a direct citation (and in the references) from text which uses 'fuel poverty'.

female partner was not in paid employment or was a woman, in employment, living alone (Broek et al. 1997). Research in Germany found a differentiation in appliance ownership between the households of single men and single women. Men tended to own more appliances related to communication, for example, mobile telephones and computers, than women did, while the reverse applied for equipment associated with household chores, such as washing machines, dryers and refrigerators (Preisendoerfer 1999).

Decision-making about appliance acquisition and use were found in mixed households, at least in Germany, to be gendered. Men were mainly responsible for decisions about purchasing appliances, such as boilers and investments in thermal insulation of the home. In contrast, women were the household energy managers with responsibility for reducing energy use not only in terms of their own household responsibilities, for example, the frequency of their use of appliances, such as washing machines and dishwashers, but also in encouraging other family members to do likewise (Buko 1996 cited in Clancy and Roehr 2003: 18). A survey in Japan examined whether there were gender differences in the behaviour responses within the household related to environmental awareness (cited in Grover et al. 1999). Women were found to be more environmentally aware which was considered to make them more inclined than men to adjust their behaviour: 45% of women compared to 41% of men switched lights off in rooms not in use, 40% of women compared to 34% of men adjusted air-conditioners to moderate temperatures and women preferred to use public transport rather than cars.

Research in the US found that mothers in paid employment were juggling work and household responsibilities, such as taking children to school and sports, shopping and other household errands. The geographical distribution of these responsibilities created time pressures which often led the women to use cars, not public transport (Sarmiento, undated). Choice of transport mode has significant implications for energy consumption. A detailed national travel survey in Sweden found a strong correlation between income and energy use, in which low-income groups tended to use public transport more frequently than higher income groups resulting in a lower annual energy consumption (Grover et al. 1999). Within the same income group, women used less energy than men which was attributed not only to work patterns but also to leisure activities since women tended to pursue these activities closer to home than men.

Another study comparing the US and Europe, which has lessons for the Global South, relates to the gender dimensions of the processes of rural electrification and the difference in approaches used to introduce consumers to using electricity (Matly 2005). The US promoted rural electrification between 1930s and 1960 with an approach which included important roles assigned to women. The rural electrification utilities were encouraged to employ women as home economists who could explain to women about the services and their associated equipment that could be powered by electricity in which they could use for their household care work. Cooperatives, which were a significant part of the US rural electrification process, were required to have both wives and husbands participate in cooperative meetings. Rural electrification boards had to have women forming at least a least, a third of the board members. The situation in Western Europe was very different. Rural electrification was completed by the mid-1930s which appears to have been entirely male-dominated since electrification was viewed entirely from a technical perspective with engineers controlling decision-making. Promotion of electrical appliances was through department stores with male sales assistants selling to women clients. Indeed, in France, women had little prospect of influencing policy since they were denied entry (until the 1970s) to the 'Grandes Ecoles' system from which many senior policymakers and business executives graduated.

2.6.2 *Women Working in the Energy Sector*

A positive finding from the research study for the European Commission mentioned above, was that more women were studying STEM subjects, at least at degree level, with some Universities, such as Bielefeld and Wilhelmshaven in Germany, offering courses specially designed for young women to motivate them to study energy from engineering and economics perspectives (Clancy and Roehr 2003). On the negative side women were still not opting to work in the energy sector. The employment data for the energy sector in the OECD countries around the end of the twentieth century showed a similar pattern. For example, in Australia: women made up 20% of the workforce in the electricity, gas and water sector but occupied less than 5% of technical positions (OSW 1999 cited in Clancy and Roehr 2003: 19). In Germany, women were around 6% of technical staff with 4% in decision-making positions, and less than 1% in top management posts (Hoppenstedt-Analyse

2000). Women were concentrated in mainly administration, personnel sales, finance and catering. There then ensued research to explain why. The highly masculine image of the sector was considered a factor in women's decision-making, a concern raised by a group of Personnel Managers within the Electricity Industry involved in a European Commission funded research project. The group produced a brochure which set out the advantages of increasing the numbers of women employed in the electricity utilities with recommendations for companies, unions and sectoral organisations to make the sector more women-friendly (ENEQO undated). Canada was already well advanced with promoting diversity in its workforce. For example, the oil and gas sector when promoting diversity in the workforce took the approach, to avoid resentment and resistance, of recognising the contributions people make as capable individuals, rather than as members of designated groups (Dowse et al. 1999). Numeric quotas were found to be insufficient as a tool to improve the statistical representation of women in the workforce, so the Canadian government and major research funding agencies switched towards programmes, such as training, which encourage women's participation and promote a more diverse workforce (Clancy et al. 2001). In the US, the legal requirement for employers to produce gender-disaggregated statistics of their workforces is considered a significant factor in increasing the number of women in science and engineering academic posts.

It is not only a question of getting women into the energy sector but also one of keeping them there. Being part of a small minority at work created for some women a sense of isolation which sometimes leads them to abandoning their careers. A study in Germany identified barriers to women remaining in employment included a male-dominated culture in the workplace, traditional hierarchies and lack of transparency in promotion processes (Accenture 2002 cited in Clancy and Roehr 2003: 21). On the positive side, attitudes and perceptions change as illustrated by several of the older women working in the energy sector interviewed in the study by Clancy and her co-workers that they considered that their younger male colleagues were becoming more socially aware and did not always conform to the 'masculine culture of work' of their older male colleagues (Clancy et al. 2001).

While energy companies and utilities saw from an instrumentalist perspective that increasing the number of women in their workforce is good for business, feminists saw renewable energy from a gender-equality perspective as providing an ideal working environment that would appeal to women both in terms of the properties of renewable energy and the type of companies involved (Rübsamen 1994). Human beings share a common pool of character traits, such as, cooperative, emotional, assertive, nurturing and non-violent which result in people being seen in roles for which a set of characteristics are associated such as leaders, supporters and intellectuals. Over time societies have classified these traits into gender stereotypes of 'female' and 'male' which become assigned to physical bodies and what are considered appropriate character traits, behaviour and roles for women and men. In many societies 'female' qualities include cooperative, non-violent and sensitive, whereas traditional 'male' qualities include competitiveness, individuality and assertiveness (Nightingale 2006). Renewable energy is associated with positive outcomes for the environment and is not destructive of the ecosystem (which can be interpreted as non-violent)—hence match 'female' qualities. In the 1970s and 1980s, which marked the early days of interest in renewable energy being considered as a major source of energy, its exploitation started with small-scale, decentralised systems run by small companies, often working on a cooperative basis, considered to provide a collaborative and supportive working environment that would align with 'female' qualities. However, as renewable energy became the domain of large companies for generating electricity using large, centralised systems the sector's image changed to being more 'professional' and men took over the strategic positions. An interesting response to this 'male take over' came in Germany where three women with technical backgrounds had become disillusioned with the male-dominated unsupportive working environment they had encountered in the traditional energy sector set up a company (Windfang) operating on co-operative principles running a 450 kW wind turbine to feed into the German national grid (see Chapter 3) (Delfs 2000).

Professional networks are considered to have a number of benefits for career development, such as acting as a source of information about employment opportunities, mentoring, offering opportunities to build skills and share experiences (Forret and Dougherty 2004). The renewable energy sector was rather slow in network development with a specific focus on women's career advancement. One of the earliest professional associations for the energy sectors, was an organisation for women working in the nuclear industry, Women in Nuclear (WIN) which was established in 1989 (Clancy et al. 2001).

2.6.3 *Women Influencing and Engendering Energy Policy*

Before 2000, there was some attention as to whether energy policies reflected women's views about the nature of energy supply and use based on the assumption that women had a greater concern than men for the environment.[21] There were a small number of studies examining whether support for renewable energy technologies or opposition to nuclear energy is gendered. A nation-wide survey in the US, conducted between 1973 and 1979, found little discernible gender-differentiated preferences in energy policy except in respect of nuclear power for which women were slightly more negative than men (Farhar et al. 1980). A similar conclusion was reached in a study held 10 years later (although based on a different cohort) (Longstreth et al. 1989).

However, a survey in Sweden in the late 1990s to gauge attitudes to the long-term use of nuclear power found 60% of young men in favour and 80% of young women were against (Puranen 2000). The economics was a driver for men and women who were in favour since nuclear power was considered good economic value for Sweden. The safety aspect of nuclear power was the major concern for the women opposed to nuclear power. The US study had a similar finding with women, more than men, prioritising the environmental quality of the energy source even where jobs were at stake (Longstreth et al. 1989). The pro-nuclear men considered nuclear as more environmentally beneficial than coal and oil with 63% of those asked having no objection to nuclear waste being stored in the municipality where they lived.

A survey in Japan of consumer awareness of environmental problems showed that at least in people over 40, women had a stronger sense than men of environmental awareness (cited in Grover et al. 1999). There were similar findings in Germany (Preisendoerfer 1999).

[21] The general assumption that women are more environmentally aware than men seems to have its origins in the appearance on the global stage of the Chipko movement in which indigenous women in India protected trees from the threat of massive logging operations (Shiva 1988). Feminists have both supported and opposed this generalisation (Resurrección 2013). Some consider that women have a natural or essential connection with nature which gives them an innate understanding of ecosystems and the environment (see, for example, Shiva 1988). Other feminists argue that it is material practices that endow women with their knowledge of ecosystems and the natural world (see, for example, Agarwal 1992). Both perspectives can be criticised for their lack of context and treating women and men as two homogeneous groups (Nightingale 2006).

Researchers suggested that levels of investment in renewable energy companies could also be an indicator of women's support for environmental protection. However, the available data showed that, at least in Europe, women only made up between 10 and 20% (on a per capita basis) of the investors in renewable energy power plants (Weller 2003 cited in Clancy et al. 2004: 23). Possibly not a surprising finding given women's relative earnings to men's and a lower likelihood of surplus funds for stock market investment.

There were attempts to influence the direction of energy policy to reflect the views of women.

Within, the International Solar Energy Society (ISES) a Women's Energy Interest Group was established in 1995 by women and men from the Global South and Global North with strong support from the ISES Board which was all male (Clancy et al. 2001). The Group's formation was in response to the perception of an overly male-dominated decision-making about energy supply. In the context of the North, it was felt that this resulted in a lack of awareness of the impact of decision about the energy mix on women's lives. Young women found participation in the meetings helped them understand the complexity of energy decisions making. The Group had ceased to function by 2004 which was attributed to a lack of a critical mass of women with time to run the group.

Women were also not influencing the research agenda. Women were underrepresented not only in absolute terms but also proportionally, for example, in the UK in 1994/1995 there were 7 women professors of engineering compared to 886 men (European Commission 2000). However, when European scientists were asked about energy research priorities, women did show a strong preference for renewable energy and energy conservation. Women researchers also supported the inclusion of social issues in energy research (Clancy et al. 2001).

2.7 Concluding Remarks

This chapter has given an overview of how the issues addressed within the gender and energy nexus have evolved since the end of the Second World War in 1945. It has described how the focus of policymakers in the energy sector has expanded from a supply-side issue of improving energy supply to include a demand-side perspective which looks at households. It could be argued that it was actually a supply-side issue, a shortage of fuelwood, which helped draw attention to the problems caused by the lack of energy in rural households of the Global South. Initially household energy was equated with meeting cooking needs which was universally identified as the responsibility of women. Searching for fuelwood was costing women a lot of time, in response to which there were well-meaning interventions designed to help women by planting more trees without taking into account the social relations of the context. Ultimately it is men, in many societies, who decide the fate of trees planted on their land. The wood from those trees was unlikely to end up under the household cooking pot. There seems to have been a lack of engagement with ethnographers who could have pointed this out.

However, researchers (see, for example, Clancy 1998; Klingshirn 2000) challenged this narrow framing and argued for a definition that encompasses all the activities that take place within a household extending beyond the practical needs to provide for the family's welfare. Cecelski (1987) had pointed out that the family home is the site not only of food and shelter but also the centre of production particularly for women. The emergence of the concept of energy services helped broaden the focus on energy supply to provide an understanding of how households perceived their needs and priorities which influence their choices of the type of energy forms they wanted. Researchers from academia and practice pointed out that these needs, priorities and capacity to make choices were differentiated across family members depending on gender, age and position in the family hierarchy. The significance of this proposition is that it brings men into the picture and their role in household energy. Evidence from different communities pointed to the male members of the household deciding about what to buy, including stoves, and who owned it (see for example, Dutta 1997; Makan 1995). Men also had different priorities to women in terms of household energy services. For men leisure was a high priority whereas for women it was a reduction in their workload. It was therefore no surprise that researchers found that the acquisition of recreational equipment, such as TVs and radios, was prioritised over labour-saving equipment. Treatment of the household as a unified entity by economists began to be challenged.

Researchers challenged the position that energy for cooking might not be the priority issue for women. Daily water collection could be more time and metabolic energy consuming than fuelwood (Mehretu and Mutambira 1992). The energy requirements of these activities also need to be considered. It was pointed out that if you don't ask women what their problems and priorities are and involve them in designing the solutions, then projects run the risk of failing, as had been seen in the low uptake in some stoves projects (Cecelski 2004). Much effort was put into trying to understand the reluctance of project staff to engage with end-users and to understand their social context. In development projects women tended to be treated as a homogeneous group. Again, energy researchers and practice, drawing on the developments in the gender and development discourse, pointed out that women are differentiated across a range of social and economic characteristics which influence, among others, capacity to act and the outcomes of interventions. Arguments were made promoting the benefits of taking a gender approach in energy project planning (see, for example, Skutsch 1998) and training materials were designed to help planners incorporate gender into their work (see, for example, Skutsch et al. 2004).

Possibly one of the most significant advances in development practice, in the period reviewed in this chapter, was the recognition that women were more than passive beneficiaries of improved energy access and upgraded technologies. Women, as users of technology, had a role to play in design. When this role was formalised into being part of a project team, they could also build skills. The next logical step was to recognise that women did run their own enterprises and so could run businesses selling, installing and maintaining energy technologies, such as PV panels.

Interest in gender issues related to energy in the Global North emerged much latter than in Global South. There have been two main themes: energy access related to social equality and the employment of women in the energy sector. Initially, social equality issues were analysed using gender-neutral terms, such as vulnerable consumers. Again, there was no inter- and intra-household differentiation with decision-makers in the energy sector reluctant to see gender as an issue for consideration. Researchers have been making a strong case for presenting data intersectionality in understanding users. Advocacy around women working in the energy sector has been related to ensuring a diverse workforce. A substantial body of research has been built on understanding why young women do not opt to study STEM subjects, if they do why they do not opt to work in the energy sector and if they do why they do not stay.

Overlaps with issues between the Global South and the Global North are beginning to emerge particularly with a wider use of the concept of energy poverty. Initially, in the Global South, attention was focused on the rural areas since supply-side issues dominated, whereas in the

Global North it was the emergence of energy poverty in urban households. However, as we shall see, urbanisation in the South has drawn attention to energy access particularly for the urban poor who may live in precarious settings, which is the situation in the Global North. In the North attention is switching to rural areas related to the use of fuelwood and peat which have become an issue related to climate change. For users, these fuels are linked to cultural identity illustrating the point found in the Global South that the other criteria than price which influence choice.

Much of what has been discussed in this chapter will reappear throughout the rest of the book.

References

Accenture. (2002). *Women and Energy. Perspectives for Female Executives in the Energy Economy*. www.strom.de

Acharya, M. and Bennett, L. (1981). *The Rural Women of Nepal: An Aggregate Analysis and Summary of Eight Village Studies.* Kathmandu, Nepal: Center for Economic Development and Administration, Status of Women Project.

Agarwal, B. (1986). *Cold Hearths and Barren Slopes: The Woodfuel Crisis in the Third World.* London: Zed Books; Delhi: Allied Publishers; and Maryland: Riverdale Publishers.

Agarwal, B. (1992). The Gender and Environment Debate: Lessons from India. *Feminist Studies* 18: 119–158.

Agarwal, B. (1997). “Bargaining” and Gender Relations: Within and Beyond the Household. *Feminist Economics* 3 (1): 1–51.

Amis, P. (2002). *Thinking About Chronic Urban Poverty*. CPRC Working Paper No 12. Chronic Poverty Research Centre, University of Manchester. ISBN 1-904049-11-7.

Angelou, N., Suh, C., Schomer, I., Marin, A.B. and Palmisano, G. (2019). *Gender and Energy Role Playing: Training Guidebook.* Washington, DC: The World Bank.

Annecke, W.J. (1999). *Concept Paper for Energy and Women: Lessons Learned.* Cape Town: EDRC, University of Cape Town.

Arnold, M., Köhlin, G., Persson, R. and Shepherd, G. (2003). *Fuelwood Revisited: What Has Changed in the Last Decade?* CIFOR Occasional Paper No. 39. Jakarta, Indonesia: Center for International Forestry Research (CIFOR).

Barnes, D. (1995). *Consequences of Energy Policies for the Urban Poor.* FPD Energy Note No 7. Washington, DC: The World Bank.

Barnes, D. and Floor, W. (1996). Rural Energy in Developing Countries: A Challenge for Economic Development. *Annual Review of Energy and the Environment* 21: 497–530.

Batliwala S. (1993). *Empowerment of Women in South Asia: Concepts and Practices*. Mumbai: Asian-South Pacific Bureau of Adult Education.

Bensch, G., Peters, J. and Schmidt, C.M. (2012). Impact Evaluation of Productive Use: An Implementation Guideline for Electrification Projects. *Energy Policy* 40: 186–195.

Bhogle, S. (2003). Rural Women as Agents of Improved Woodstove Dissemination: A Case-Study in Huluvangala Village, Karnataka, India. *ESD* VII (3): 70–75.

Boardman, B. (1998). *Energy Efficiency and Fuel Poverty*. PRASEG Annual Conference, 5 November 1998.

Boserup, E. (1970). *Women's Role in Economic Development*. London: Allen & Unwin.

Bouzarovski, S. and Petrova, S. (2015). A Global Perspective on Domestic Energy Deprivation: Overcoming the Energy Poverty–Fuel Poverty Binary. *ERSS* 10: 31–40.

Broek, A. van den, Knulst, W.P. and Niggebrugge, F.D.E. (1997). *Trends in de tijdsbesteding van de Nederlandse bevolking 1975–1995*. Werkdocument 41, Sociaal Cultureel Planbureau (SCP).

Buko. (1996). *Zwischen sparstruempfen und gigabytes -- der oekologen lust, der frauen frust: Eine feministische kritik zur studie 'Nachhaltiges Deutschland*. In Forum entwicklungspolitischer Aktionsgruppen, Heft 201/96 (Feminist Critiques on the Study "Sustainable Germany").

Buzar, S. (2007). The 'Hidden' Geographies of Energy Poverty in Post-socialism: Between Institutions and Households. *Geoforum* 38: 224–240.

Catalyst. (2023). *Male-Dominated Occupations Are Those Held Primarily by Men Rather Than Women*. https://www.catalyst.org/research/women-in-male-dominated-industries-and-occupations/ (accessed 7 June 2023).

Cecelski, E. (1987). Energy and Rural Women's Work: Crisis, Response and Policy Alternatives. *International Labour Review* 126 (I): 41–64.

Cecelski, E. (1995). From Rio to Beijing: Engendering the Energy Debate. *Energy Policy* 23 (6): 561–575.

Cecelski, E. (2004). *Re-thinking Gender and Energy: Old and New Directions*. ENERGIA/EASE Discussion Paper. www.energia.org.

Cecelski, E. (2005). *Energy, Development and Gender: Global Correlations and Causality*. Report Prepared as part of the United Kingdom Department for International Development (DfID) KaR research project R8346 on "Gender as a Key Variable in Energy Interventions." www.energia.org (accessed 8 March 2023).

Cecelski, E. and Dutta, S. (2011). *Mainstreaming Gender in Energy Projects: A Practical Handbook*. ENERGIA. www.energia.org

Cecelski, E. and Matinga, M. (2014). *Cooking With Gas: Why Women in Developing Countries Want LPG and How They Can Get It*. Report Developed by ENERGIA for the World LP Gas Association.

Chambers, R. and Conway, G. (1992). *Sustainable Rural Livelihoods: Practical Concepts for the 21st Century*. Brighton, UK: Institute for Development Studies.

Childs, S. and Krook, M. L. (2009). Analysing Women's Substantive Representation: From Critical Mass to Critical Actors. *Government and Opposition* 44 (2): 125–145.

Christian Michelsen Institute. (1999). WID/Gender Units and the Experience of Gender Mainstreaming in Multilateral Organisations: Knights on White Horses? Evaluation Report 1.99 for the Royal Ministry of Foreign Affairs, Oslo: Christian Michelsen Institute.

Clancy, J.S. (1998). Household Energy Isn't All Stoves. *Boiling Point* No 41. London: IT Publications.

Clancy, J.S. (2002). *Blowing the Smoke Out of the Kitchen: Gender Issues in Household Energy*. SPARKNET Briefing Paper 2. https://research.utwente.nl/en/publications/blowing-the-smoke-out-of-the-kitchen-core-issues-in-household-ene (accessed 24 February 2022).

Clancy, J. (2006). *Enabling Urban Poor Livelihoods Policy Making: Understanding the Role of Energy Services*. Briefing Paper Prepared for DFID KaR Project R8348. https://www.gov.uk/research-for-development-outputs/urban-poor-livelihoods-understanding-the-role-of-energy-services (accessed 11 January 2023).

Clancy, J. and Feenstra, M. (2006). *How to Engender Energy Policy*. A Technical Briefing Paper Prepared for ENERGIA. The Hague: ENERGIA/HIVOS.

Clancy, J.S., Maduka, O. and Lumampao, F. (2006). *Enabling Urban Poor Livelihoods Policy Making: Understanding the Role of Energy Services*. Synthesis Report Prepared for DFID KaR R8348.

Clancy, J.S., Maduka, O. and Lumampao, F. (2007). *Sustainable Energy Systems and Urban Poor Livelihoods*. In Droege, P. (ed.), Urban Energy Transition. Elsevier.

Clancy, J.S. and Roehr, U. (2003). Gender and Energy: Is There a Northern Perspective? *J. Energy for Sustainable Development* VII (3): 44–50.

Clancy, J.S., Skutsch, M.M. and Batchelor, S. (2003). *The Gender-Energy-Poverty Nexus: Can We Find the Energy to Address Gender Concerns in Development?* DFID Project CNTR998521.

Clancy, J., Oparaocha, S. and Roehr, U. (2004). Gender Equity and Renewable Energies. Thematic Background Paper. *International Conference for Renewable Energies*. Bonn.

Clancy, J.S., Stienstra, M., Gregory, J. and Cornland, D. (2001). *Gender and Energy–Women's Concerns in Energy: Background and State of the Art.*

Working Paper 1. Study Carried Out for the European Commission Research Directorate.

Clancy, J.S., Visser, J. and Kamphuis, E. (2009). *The Gender Face of Energy*. ENERGIA EU-COOPENER Project.

Cornwall, A. (2016). Women's Empowerment: What Works? *Journal of International Development* 28: 342–359.

Cornwall, A. and Rivas, A.-M. (2015). From 'Gender Equality and 'Women's Empowerment' to Global Justice: Reclaiming a Transformative Agenda for Gender and Development. *Third World Quarterly* 36 (2): 396–415.

Cornwall, A. and White, S.C. (2000). Men, Masculinities and Development: Politics, Policies and Practice. *IDS Bulletin* 31 (2): 1–6. Brighton, UK: Institute of Development Studies.

Cornwall, A. Harrison, E. and Whitehead, A. (2004). Introduction to Repositioning Feminisms in Gender and Development. *IDS Bulletin* 35 (4): 1–10. Brighton, UK: Institute of Development Studies.

Council of Europe. (2014). *Council of Europe Gender Equality Strategy 2014–2017*. Strasbourg: Council of Europe.

Crenshaw, K., Gotanda, N., Peller, G. and Kendall, T. (1995). *Critical Race Theory: The Key Writings That Formed the Movement*. New York: The New Press.

CRGGE (ENERGIA/DfID Collaborative Research Group on Gender and Energy). (2006). *From the Millennium Development Goals Towards a Gender-Sensitive Energy Policy Research and Practice: Empirical Evidence and Case Studies*. Synthesis Report. UK Department for International Development (DfID) KaR research project R8346 "Gender as a Key Variable in Energy Interventions: Are We Asking the Right Questions". https://www.energia.org/publications/from-the-millenium-development-goals-towards-a-gender-sensitive-energy-policy-research-and-practice-empirical-evidence-and-case-studies/ (accessed 24 February 2022).

Dalaker, J. (2001). *Poverty in the United States: 2000*. U.S. Census Bureau, Current Population Reports, Series P60-214. Washington, DC: U.S. Government Printing Office.

Daly, M. (2005). Gender Mainstreaming in Theory and Practice. *Social Politics: International Studies in Gender, State & Society* 12 (3): 433–445.

Davis, K. (2008). Intersectionality as Buzzword: A Sociology of Science Perspective on What Makes a Feminist Theory Successful. *Feminist Theory* 9 (1): 67–85.

De Jong, S. (2016). Mainstreaming Has Never Run Clean, Perhaps Never Can: Gender in the Main/Stream of Development. In Harcourt, W. (ed.), *The Palgrave Handbook of Gender and Development*. 2016. Basingstoke, UK: Palgrave Macmillan.

Delfs, C. (2000). Case-Study: Women Producing Electricity from Renewables: From 'Project' to 'Market': The 'Windfang' Project in Germany. *ENERGIA News* 3 (1).

D'Enbeau, S. (2011). Transnational Feminist Advocacy Online: Identity (Re)Creation Through Diversity, Transparency, and Co-Construction. *Women's Studies in Communication* 34 (1): 64–83.

Derbyshire, H. (2002). *Gender Manual: A Practical Guide for Development Policy Makers and Practitioners*. London: Social Development Division, DfID (Department for International Development).

Dowse, S., Horton, K., Lele D. and Sherk, S. (1999). *Women in Canada's Oil and Gas Sector*. Report Prepared for the Canadian International Development Agency's Oil and Gas Sector Programme Pakistan.

Dutta, S. (1997). Role of Women in Rural Energy Programmes: Issues, Problems and Opportunities. *ENERGIA News* 4: 11–14.

Dutta, S. (2003a). *Energy as a Key Variable in Eradicating Extreme Poverty and Hunger: A Gender and Energy Perspective on Empirical Evidence on MDG #1*. Discussion Paper Prepared for DFID/ENERGIA Project on Gender as a Key Variable in Energy Interventions.

Dutta, S. (2003b). *Guidelines on Engendering Energy for Rural Development*. Presentation to Review Meeting on Integration of Energy and Rural Development Policies and Programmes, UN Economic Commission for Asia and the Pacific, Bangkok, 25–26 September.

Eckholm, E. (1975). *The Other Energy Crisis: Firewood*. Worldwatch Paper No. 1. Washington, DC: World Resources Institute.

Elson, D. (2012). Review of World Development Report 2012. *Global Social Policy* 12 (2): 178–183

ENERGIA. (2007). *The Gender Face of Energy*. Training Manual. www.energi a.org (accessed 1 March 2023).

ENERGIA Support Group and CSD-NGO Women's Caucus. (2001). *Gender Perspectives on Energy for CSD-9*. Position Paper Prepared for the 9th Session of the Commission on Sustainable Development. http://www.energia.org (accessed 1 March 2023).

ENEQO. (undated). *Equal Opportunities in the Electricity Sector: How to Get Better Value From Your People Resources While Providing Good Careers for Women and Men*. Brochure produced as part of the Equal Opportunities in the Energy Sector (ENEQO) project funded under the European Community Fourth Action Programme on Equal Opportunities for Women and Men.

Escobar, A. (1995). *Encountering Development: The Making and Unmaking of the Third World*. Princetown, NJ, USA: Princetown University Press.

Escobar, A. (1998). Whose Knowledge, Whose Nature? Biodiversity, Conservation, and the Political Ecology of Social Movements. *Journal of Political Ecology* 5: 53–82.

ESMAP. (1999). *Household Energy Strategies for Urban India: The Case of Hyderabad*. Washington, DC: World Bank.

ESMAP. (2020). *Cooking With Electricity: A Cost Perspective*. Washington, DC: World Bank.

Esquivel, V. and Sweetman, C. (2016). Gender and the Sustainable Development Goals. *Gender & Development* 24 (1): 1–8.

European Commission. (2000). *Science Policies in the European Union: Promoting Excellence Through Mainstreaming Gender Equality*. A report from the ETAN Expert Working Group on Women and Science. Luxembourg: Office for Official Publications of the European Communities.

Everts, S. and Schulte, B. (1997). Vietnam Women's Union Promotes Solar Energy. *ENERGIA News* (3): 12–13.

Farhar, B.C., Unseld C.T., Vories, R. and Crews, R. (1980). Public Opinion About Energy. *Annual Review of Energy* 5: 141–172.

Feenstra, M. (2021). *Gender Just Energy Policy: Engendering the Energy Transition in Europe*. PhD Thesis. University of Twente, The Netherlands.

Forret, M. and Dougherty, T. (2004). Networking Behaviors and Career Outcomes: Differences for Men and Women? *Journal of Organizational Behavior* 25: 419–437.

Foster, V. and Rana, A. (2020). *Rethinking Power Sector Reform in the Developing World*. Washington, DC: The World Bank.

Friedmann, J. (1992). *Empowerment. The Politics of Alternative Development*. Cambridge, UK: Blackwell.

Goetz, A-M. (2007). *Gender Justice, Citizenship and Entitlements—Core Concepts, Central Debates and New Directions for Research*. In Mukhopadhyay, M. and Singh, N. (eds.) Gender Justice, Citizenship and Development (pp. 15–57). Ottawa: International Development Research Centre.

Goldman, M.J. and Little, J.S. (2015). Innovative Grassroots NGOs and the Complex Processes of Women's Empowerment: An Empirical Investigation from Northern Tanzania. *World Development* 66: 762–777.

Grover, S., Flenley, C. and Hemmati, M. (1999). *Gender and Sustainable Consumption: Bridging Policy Gaps in the Context of Chapter 4, Agenda 21—Changing Consumption and Production Patterns*. Report for the United Nations Commission on Sustainable Development, 7th Session, April 1999 (accessed 1 March 2023).

Havet, I., Braun, F. and Gocht, B. (2007). *Gender Mainstreaming: A Key Driver of Development in Environment & Energy*. Training Manual Prepared for UNDP. New York: United Nations Development Programme (UNDP).

Hearn, J. and Morgan, D.H.J. (2014). *Men, Masculinities and Social Theory*. Routledge.

Hirtenfelder, C.T. (ed.) (2015). *A Research Framework on Gender and Socio-Cultural Inclusion in Global Change Research.* Report Prepared for the Human Sciences Research Council, South Africa.

Hoppenstedt-Analyse. (2000). *Frauen im Management 1999* (Women in Management 1999). Branchenanalyse 1, Hoppenstedt-Verlag.

Hospes, O. and Clancy, J.S. (2011). *Unpacking the Discourse of Social Inclusion in Value Chains.* In Helmsing, A.H.J. and Vellema, S. (eds.), Value Chains, Inclusion and Endogenous Development: Contrasting Theories and Realities. Routledge.

Howard-Hassmann, R.E. (2011). Universal Women's Rights Since 1970: The Centrality of Autonomy and Agency. *Journal of Human Rights* 10: 433–449.

IDS. (2003). *Energy, Poverty and Gender: A Review of the Evidence and Case Studies in Rural China.* Report Prepared for the World Bank by the Institute for Development Studies (IDS). Washington, DC: The World Bank.

IISD, BIDS, IRADe and Spaces for Change. (2019). *Gender and Fossil Fuel Subsidy Reform: Findings From and Recommendations for Bangladesh, India and Nigeria.* Report Prepared for the ENERGIA Gender and Energy Research Programme. The Hauge: ENERGIA.

James, B. (1998). *Community Participation in Rural Electrification: Community Based Organisations for Operation, Maintenance and Administration in Rural Electrification.* Cape Town, South Africa: Energy & Development Centre, University of Cape Town.

Jones, N., Tvedten, I., Arbulú, A., Pereznieto, J., Lindström, P. and Norbakk, M. (2015). *Evaluation of Norway's Support to Women's Rights and Gender Equality in Development Cooperation.* Report Prepared for Norad Evaluation Department. Oslo: Norad (Norwegian Agency for Development Cooperation).

Joseph, T. (2009). *Mainstreaming Women in Development? A Gender Analysis of the UNDP in South Africa.* PhD Dissertation, Stellenbosch University: South Africa.

Kabeer, N. (1999). Resources, Agency, Achievements: Reflections on the Measurement of Women's Empowerment. *Development and Change* 30: 435–464.

Kabeer, N. (2000). Social Exclusion, Poverty and Discrimination: Towards an Analytical Framework. *IDS Bulletin* 31: 83–97. Brighton, UK: Institute of Development Studies.

Kabeer, N. (2005). Gender Equality and Women's Empowerment: A Critical Analysis of the Third Millennium Development Goal. *Gender and Development* 13 (1): 13–24.

Kabeer, N. (2012a). *Women's Economic Empowerment and Inclusive Growth: Labour Markets and Enterprise Development.* Ottawa, Canada: IDRC.

Kabeer, N. (2012b). Empowerment, Citizenship and Gender Justice: A Contribution to Locally Grounded Theories of Change in Women's Lives, Ethics and Social Welfare. *Ethics and Social Welfare* 6 (3): 216–232.

Kabeer, N. (2016). *Gender Equality, the MDGs and the SDGs: Achievements, Lessons and Concerns*. blogs.lse.ac.uk/southasia/2015/10/05/gender-equality-the-mdgs-and-the-sdgs-achievements-lessons-andconcerns/ (accessed 25 February 2021).

Karlsson, G. and Clancy, J.S. (2004). *Toolkit and Resource Guide: Mainstreaming Gender in Energy Management*. New York: UNDP. https://www.undp.org/publications/energy-and-gender-sustainable-development-toolkit-and-resource-guide (accessed 16 December 2022).

Kelkar, G., Nathan, D., Rengalakshmi, R., Manjula, M., Gaikwad, S., Shakya, I., Shrestha, P. and Tamang, A.P. (2019). *The Gender Factor in Political Economy of Energy Sector Dynamics*. Scoping Study Report Prepared for the ENERGIA Gender and Energy Research Programme. The Hauge: ENERGIA.

Khamati Njenga, B. and Clancy, J. (2002). *Concepts and Issues in Gender and Energy*. ENERGIA. https://research.utwente.nl/en/publications/concepts-and-issues-in-gender-and-energy (accessed 16 December 2022).

Klingshirn, A. (2000). *Integrating Household Energy into Rural Development Programmes*. https://www.energia.org/publications/integrating-household-energy-into-rural-development-programme/ (accessed 24 February 2022).

Kojima, M. (2011). *The Role of Liquified Petroleum Gas in Reducing Energy Poverty*. Extractive Industries for Development Series 25. Washington, DC: The World Bank.

Leach, G. and Mearns, R. (1988). *Beyond the Woodfuel Crisis: People, Land and Trees in Africa*. London: Earthscan.

Lee, A.D. and Usman, Z. (2018). *Taking Stock of the Political Economy of Power Sector Reforms in Developing Countries: A Literature Review*. Policy Research Working Paper 8518. Washington, DC: The World Bank.

Legros, G., Kamal, R. and Seyedi, B. (2011). *Decentralised Energy Access and the Millennium Development Goals. An Analysis of the Development Benefits of Micro-Hydropower in Rural Nepal*. United Nations Development Programme, Alternative Energy Promotion Centre and Practical Action.

Listo, R. (2018). Gender Myths in Energy Poverty Literature: A Critical Discourse Analysis. *Energy Research & Social Science* 38: 9–18.

Longstreth, M., Turner, J., Topliff, M. and Iams, D. (1989). Support for Soft and Hard Path American Energy Policies: Does Gender Play a Role? *Women's Studies International Forum* 12 (2): 213–226.

Lugones, M. (2016). The Coloniality of Gender. In Harcourt, W. (ed.), *The Palgrave Handbook of Gender and Development*. Basingstoke, UK: Palgrave Macmillan.

Makan, A. (1995). Power for Women and Men: Towards a Gendered Approach to Domestic Policy and Planning in South Africa. *Third World Policy Review* 17 (2): 183–198.

Martinot, E., Chaurey, A., Lew, D., Moreira, J.R. and Wamukonya, N. (2002). Renewable Energy Markets in Developing Countries. *Annual Review of Energy and the Environment* 27: 309–48.

Masera, O.R., Bailis, R., Drigo, R., Ghilardi, A. and Ruiz-Mercado, I. (2015). Environmental Burden of Traditional Bioenergy Use. *Annual Review of Environment and Resources* 40: 121–150.

Masera, O.R., Saatkamp, B.D. and Kammen, D.M. (2000). From Linear Fuel Switching to Multiple Cooking Strategies: A Critique and Alternative to the Energy Ladder Model. *World Development* 28 (12): 2083–2103.

Matinga, M. and Clancy, J. (2020). *Gender, Firewood and Health: The Potential of Ethnography to Inform Policy and Practice*. In Clancy, J., Őzeröl, G., Mohlakoana, N., Feenstra, M. and Sol Cueva, L. (eds.), Engendering the Energy Transition. Basingstoke, UK: Palgrave Macmillan.

Matly, M. (2005). *Women's Electrification*. Report Prepared for the United Kingdom Department for International Development (DFID) KaR Research Project R8346 on "Gender as a Key Variable in Energy Interventions." https://www.gov.uk/research-for-development-outputs/women-s-electrification.

McDade, S. and Clancy, J. (2003). Editorial. *Energy for Sustainable Development* 7 (3): 3–7.

Mehretu, A. and Mutambira, C. (1992). Gender Differences in Time and Energy Costs of Distance for Regular Domestic Chores in Rural Zimbabwe. *World Development* 20 (11): 1675–1683.

Middleton, E. (2010). A Political Ecology of Healing. *Journal of Political Ecology* 17: 1–28.

Modi, V. (2004). *Energy Services for the Poor*. Commissioned Paper for the Millennium Project Task Force 1, Draft November 9, 2004, Earth Institute and Department of Mechanical Engineering, Columbia University, New York.

Moghadam, V.M. (2000). Transnational Feminist Networks: Collective Action in an Era of Globalization. *International Sociology* 15 (1): 57–85.

Molyneux, M. (1985). Mobilization Without Emancipation? Women's Interests, State and Revolution in Nicaragua. *Feminist Studies* 11 (2): 227–254.

Molyneux, M. (2004). The Chimera of Success. *IDS Bulletin* 35 (4): 112–116.

Moser, C. (1989). Gender Planning in the Third World: Meeting Practical and Strategic Gender Needs. *World Development* 17 (11): 1799–1825.

Moser, C. (1993). *Gender Planning and Development: Theory, Practice and Training*. London: Routledge.

Moser, C. (2005). Has Gender Mainstreaming Failed? *International Feminist Journal of Politics* 7 (4): 576–590.

MSSRF and CRT Nepal. (2019). *The Gender Factor in Political Economy of Energy Sector Dynamics*. Research Report RA3 Prepared for the ENERGIA Gender and Energy Research Programme 'Building the Evidence Base for Improving Energy Interventions' Effectiveness by Taking'. www.energia.org. The Hague: ENERGIA/HIVOS.

Mukhopadhyay, M. (2016). Mainstreaming Gender or "Streaming" Gender Away: Feminists Marooned in the Development Business. In Harcourt, W. (ed.), *The Palgrave Handbook of Gender and Development*. Basingstoke, UK: Palgrave Macmillan.

Narayan, D. (1999). *Can Anyone Hear Us? Voices From 47 Countries*. Washington, DC: The World Bank.

Nightingale, A. (2006). The Nature of Gender: Work, Gender, and Environment. *Environment and Planning D: Society and Space* 24: 165–185.

Njenga, B.K. (2001). *Upesi Rural Stoves Project*. In Karlsson, G.V. (ed.), Generating Opportunities: Case Studies on Energy and Women. New York: UNDP.

Nussbaum, M. (2000). Women's Capabilities and Social Justice. *Journal of Human Development* 1 (2): 219–247.

Office of the Status of Women (OSW). (1999). *Women in Australia 99*. Canberra: Commonwealth Government of Australia.

Ogundipe-Leslie, O. (1994). *Re-creating Ourselves: African Women & Critical Transformations*. Trenton, NJ: Africa World Press.

O'Neil, T. and Domingo, P. (2016). *Women and Power: Overcoming Barriers to Leadership and Influence*. London: Overseas Development Institute (ODI).

Overholt, C.A., Could, K., Anderson, M.B. and Austin, J.E. (1985). *Gender Roles in Development Projects: A Case Book*. West Hartford: Kumarian Press.

Oyěwùmí, O. (1997). *The Invention of Women: Making an African Sense of Western Gender Discourses*. Minneapolis: University of Minnesota Press.

Pachauri, S. and Spreng, D. (2011). Measuring and Monitoring Energy Poverty. *Energy Policy* 39 (12): 7497–7504.

Pachauri, S., Scott, A., Scott, L. and Shepherd, A. (2013). *Energy for All: Harnessing the Power of Energy Access for Chronic Poverty Reduction*. Energy Policy Guide No. 3. London, UK: Chronic Poverty Advisory Network.

Parpart, J.L. (2014). Exploring the Transformative Potential of Gender Mainstreaming in International Development Institutions. *Journal of International Development* 26: 382–395

Parpart, J.L., Connelly, M.P. and Barriteau, V.E. (eds.). (2000). *Theoretical Perspectives on Gender and Development*. Ottawa, Canada: IDRC.

Parpart, J.L., Rai, S.M. and Staudt, K. (2002). Rethinking Em(power)ment, Gender and Development: An Introduction. In J.L. Parpart, S.M. Rai and K. A. Staudt (eds.), *Rethinking Empowerment: Gender and Development in a Global/Local World* (pp. 3–21). New York: Routledge.

Paulson, S. (2016). Towards a Broader Scope and More Critical Frame for Intersectional Analysis. In Harcourt, W. (ed.), *The Palgrave Handbook of Gender and Development*. Basingstoke, UK: Palgrave Macmillan.

Pearson, R. (2004). Women, Work and Empowerment in a Global Era. *IDS Bulletin* 35 (4): 117–120.

Petroleum Planning & Analysis Cell (PPAC). (2018). *LPG Profile (Data on LPG Marketing) as on 01.04.2018*. Petroleum Planning & Analysis Cell, Ministry of Petroleum & Natural Gas. Retrieved from http://ppac.org.in/WriteReadData/Reports/201805290110218094341DataonLPGMarketing01042018.pdf

Preisendoerfer, P. (1999). *Umwelteinstellungen und Umweltverhalten in Deutschland: Empirische Befunde und Analysen auf der Grundlage der Bevoelkerungsumfragen "Umweltbewusstsein in Deutschland 1991–1998"*. Herausgegeben vom Umweltbundesamt (Environmental Attitudes and Environmental Behaviour in Germany), Leske und Budrich, Opladen.

Pulé, P. M. and Hultman, M. (2021). *Men, Masculinities, and Earth: Contending With the (m)Anthropocene*. London: Palgrave.

Puranen, B. (2000). *Young Women and Men: Increasingly Alike?* http://www.bikupan.se/young/ (accessed 7 March 2023).

Quijano, A. (2000). Colonialidad del poder y clasificación social. *Journal of World System Research* 1 (3): 342–388.

Ramani, K.V. and Heijndermans, E. (2003). *Energy, Poverty and Gender: A Synthesis*. Washington, DC: The World Bank.

Rao, N.D. (2013). Does (Better) Electricity Supply Increase Household Enterprise Income in India? *Energy Policy* 57: 532–541.

Rao, A. and Sandler, J. (2016). Beyond Binaries: Strategies for a 21st Century Gender Equality Agenda. In Harcourt, W. (ed.), *The Palgrave Handbook of Gender and Development*. Basingstoke, UK: Palgrave Macmillan.

Rathgeber, E.M. (1990). WID, WAD, GAD: Trends in Research and Practice. *Journal of Development Areas* 24 (4): 489–502.

Razavi, S. (2016). The 2030 Agenda: Challenges of Implementation to Attain Gender Equality and Women's Rights. *Gender & Development* 24 (1): 25–41.

Reddy, A.K.N. (2000). Energy and Social Issues. In: Goldemberg, J. (ed.), *World Energy Assessment: Energy and the Challenge of Sustainability*. New York: UNDP.

Resurrección, B.P. (2013). Persistent Women and Environment Linkages in Climate Change and Sustainable Development Agendas. *Women's Studies International Forum* 40: 33–43.

Rübsamen, R. (1994). *Alternativenergie wird erwachsen—Waren Frauen nur für die Kinderstube gut?* In Buchen, J., Buchholz, K., Hoffmann, E., Hofmeister, S., Kutzner, R., Olbrich R., und van Rüth, P. (eds.), Das Umweltproblem ist nicht geschlechtsneutral—Feministische Perspektiven. Bielefeld.

Sainsbury, D. and Bergqvist, C. (2009). The Promise and Pitfalls of Gender Mainstreaming. *International Feminist Journal of Politics* 11 (2): 216–234.

Saith, A. (2006). From Universal Values to Millennium Development Goals: Lost in Translation. *Development and Change* 37 (6): 1167–1199.

Sarmiento, S. (undated). *Household, Gender and Travel.* www.fhwa.dot.gov/ohim/womens/chap3.pdf (accessed 1 March 2023).

Scott, C.V. (1995). *Gender and Development: Rethinking Modernization and Dependency Theory*. Boulder, USA: Lynne Rienner Publishers.

Sen, A. (1999). *Development as Freedom.* New York: Random House.

Sen, G. (1997). *Empowerment as an Approach to Poverty.* Working Paper Series 97.07, Background Paper for the UNDP Human Development Report. New York: UNDP.

Sen, G. and Grown, C. (1987). *Development Crises and Alternative Visions: Third World Women's Perspectives.* London: Earthscan.

Sen, G. and Mukherjee, A. (2014). No Empowerment Without Rights, No Rights Without Politics: Gender-Equality, MDGs and the Post-2015 Development Agenda. *Journal of Human Development and Capabilities* 15: 188–202.

Sharma, B., Banskota, K. and Luitel, S. (2005). *Women, Energy and Water in the Himalayas.* Training Manual. Nairobi, Kenya: United Nations Environment Programme (UNEP) and Kathmandu, Nepal: International Centre for Integrated Mountain Development.

Shiva, V. (1988). *Staying Alive: Women, Ecology and Survival in India.* London: Zed Books.

Sinha, S. (2012). *In Pursuit of a Light Bulb and a Smokeless Kitchen.* PhD Thesis, University of Twente, The Netherlands.

Skutsch, M.M. (n.d.). Tooling Up for Gender and Energy. Paper Prepared for ENERGIA. https://www.energia.org/assets/2015/06/77-Tooling-Up-For-Gender-and-Energy.pdf (accessed 12 December 2022).

Skutsch, M.M. (1998). The Gender Issue in Energy Project Planning: Welfare, Empowerment or Efficiency? *Energy Policy* 26 (12): 945–955.

Skutsch, M.M. (2005). Gender Analysis for Energy Projects and Programmes. *ESD* IX (1): 37–52.

Skutsch, M.M., Clancy, J.S. and Leeuw, H. (2004). *The Gender Face of Energy: A Training Manual.* Leusden, the Netherlands: ENERGIA.

Skutsch, M.M., van der Molen, I. and Clancy, J.S. (1999). *Equity Versus Efficiency: Gender Versus Participation—Contest in the Area of Donor Paradigms.* In Proceedings of the 9th EADI General Conference: Europe and the South in the 21st Century. Paris, September 1999.

Smith, K.R. (1999). *Indoor Air Pollution.* Pollution Management in Focus Discussion Note No. 4. August 1999. Washington, DC: World Bank.

Smith, K. (2002). In Praise of Petroleum? *Science* 298 (5600): 1847.

Sovacool, B.K. and Dworkin, M.H. (2015). Energy Justice: Conceptual Insights and Practical Applications. *Applied Energy* 142: 435–444.

Standal, K. (2008). *Giving Light and Hope in Rural Afghanistan: The Impact of Norwegian Church Aid's Barefoot Approach on Women Beneficiaries*. Master's Thesis, University of Oslo, Norway.

Stern, D.I. (2011). The Role of Energy in Economic Growth. *Annals of the New York Academy of Sciences* 1219: 26–51.

Stoecklein, S. (1997). Teenage Daughters and Other Unpredictables. *Build*, February–March: 36–38.

Stojilovska, A., Yoon, H. and Robert, C. (2021). Out of the Margins, into the Light: Exploring Energy Poverty and Household Coping Strategies in Austria, North Macedonia, France, and Spain. *Energy Research and Social Science* 82: 102279.

Tambe, A. and Trotz, A. (2010). Historical Reflections on DAWN: An Interview With Gita Sen. *Comparative Studies of South Asia, Africa and the Middle East* 30 (2): 214–217.

Tamma, S. and Duile, T. (2020). Indigeneity and the State in Indonesia: The Local Turn in the Dialectic of Recognition. *Journal of Current Southeast Asian Affairs* 39 (2): 270–289.

Tinker, I. (1987). The Real Rural Energy Crisis: Women's Time. *Energy Journal* 3: 125–146.

UNDP. (2004). *Gender and Energy for Sustainable Development: A Tool Kit and Resource Guide*. In Karlson, G. and Clancy J. (eds.). New York: United Nations Development Programme.

UNDP. (2006). *Expanding Access to Modern Energy Services: Replicating, Scaling Up and Mainstreaming at the Local Level*. New York: United Nations Development Programme.

UNDP/UNIES. (2000). *Sustainable Energy Strategies: Materials for Decision Makers*. New York: United Nations Development Programme.

UNESCO (United Nations Educational, Scientific and Cultural Organization). (2017). *Cracking the Code: Girls' and Women's Education in Science, Technology, Engineering and Mathematics (STEM)*.

Valentine, G. (2007). Theorizing and Researching Intersectionality: A Challenge for Feminist Geography. *The Professional Geographer* 59 (1): 10–21.

van Berkel, R., Møller, I.H. and Williams, C.C. (2002). The Concept of Inclusion/Exclusion and the Concept of Work. In R. van Berkel and I.H. Møller (eds.), *Active Social Policies in the EU: Inclusion Through Participation*. Bristol: The Policy Press.

Walsh, C. (2016). Gender and Its 'Otherwise'. In Harcourt, W. (ed.), *The Palgrave Handbook of Gender and Development*. Basingstoke, UK: Palgrave Macmillan.

Weller, I. (2003). *Gender Impact Assessment der angewandten Umweltforschung Bremen.* Unveröffentlichter Abschlussbericht zum Forschungsvorhaben. University of Bremen.

Wickramasinghe, A. (2003). Gender and Health Issues in the Biomass Energy Cycle: Impediments to Sustainable Development. *ESD* 7 (3): 51–61.

Wieringa, S. (1994). Women's Interests and Empowerment: Gender Planning Reconsidered. *Development and Change* 25 (4): 829–848.

Wilson, M. and Green, J.M. (2000). The Feasibility of Introducing Solar Ovens to Rural Women in Maphephethe. *Tydskrif vir Gesinsekologie en Verbruikerswetenskappe* 28: 54–61. ISSN 0378-5254

Winker, G. and Degele, N. (2011). Intersectionality as Multi-Level Analysis: Dealing With Social Inequality. *European Journal of Women's Studies* 18 (1): 51–66.

World Bank. (1996). *Rural Energy and Development: Improving Energy Supplies for Two Billion People.* Washington, DC: The World Bank.

World Bank. (2000). *Energy Services for the World's Poor.* Washington, DC: The World Bank.

World Bank. (2007). *Gender Equality as Smart Economics.* Washington, DC: The World Bank.

World Bank. (2011). *World Development Report 2012.* Washington, DC: The World Bank.

Woroniuk, B. and Schalkwyk, J. (1998). *Energy Policy and Equality Between Women and Men.* SIDA Equality Prompt #9. Stockholm: Infrastructure Division, SIDA.

Young, K. (1993). *Planning Development With Women.* London and Basingstoke, UK: Macmillan Educational.

Yuval-Davis, N. (2011). *Power, Intersectionality and the Politics of Belonging.* Working Paper No. 75, Feminist Research Center (FREIA) Working Paper Series. Aalborg, Denmark: FREIA, Aalborg University.

Yuval-Davis, N. (2016). Power, Intersectionality and Politics of Belonging. In Harcourt, W. (ed.), *The Palgrave Handbook of Gender and Development.* Basingstoke, UK: Palgrave Macmillan.

CHAPTER 3

Energy and Low-Income Households

3.1 Introduction

This chapter looks at energy use in households, particularly in respect of low-income households, in the context of the energy transition.

The chapter starts with some statistics about levels of energy access. After which the concept of energy poverty, which introduced in Chapter 2, is taken further. This section is followed by a discussion of the gender dimensions of energy poverty. The impacts of energy poverty on health, time poverty and drudgery are then described in more detail. This is followed by describing approaches to promoting energy access as a means to addressing energy poverty. A summary of the empirical evidence about what practical changes, as well as the influence on gender roles and relations, related to household access to energy services is presented before the conclusions which close the chapter.

3.2 Levels of Energy Access

The 2023 Global Tracking Framework Report, which monitors progress with SDG7, reported that, in 2021, 675 million people did not have access to electricity, 80% of whom live in Sub-Saharan Africa. This figure has almost halved since 2010 when there were 1.1 billion people without electricity access. The decrease in the numbers of people with access to clean cooking in the same time period has not been as dramatic, the

J. Clancy, *Engendering the Energy System*,
https://doi.org/10.1007/978-3-031-78965-6_3

numbers fell from 2.9 billion to 2.3 billion (World Bank 2023). However, there appear to be significant differences in levels of access clean cooking fuels linked to the location of the household. An analysis of the data from the Demographic and Health Surveys (DHS) for 31 Sub-Saharan African countries for the period 2010–2016 found that 26% of urban households had access to clean cooking fuels while 2% of rural households had access (Ato Armah et al. 2019).

There is also a difference between rural and urban access to electricity: 85% compared to 98%. Population growth in rural Sub-Saharan Africa was not matched by increased installed electricity infrastructure whereas in Asia, rural access figures improved due to urban migration. In respect of slower access to clean cooking, in rural Sub-Saharan Africa, this has been attributed to the population increasing at a faster rate. While urban areas have access figures of 86% compared to 51% in rural areas, there is some concern that the access rate is slowing and might in the not-too-distant future stop declining.

It would be incorrect to assume that all households in the Global North are connected to a modern energy supply. The dramatic increase in energy prices in the 2020s has seen many low-income households not be able to pay their energy bills. In the UK, it has been estimated that, in 2022, more than two million people were being disconnected at least once a month from their gas or electricity supply because they could not top up their pre-payment meters (Citizens Advice 2023). There are households throughout the EU, but particularly in Eastern Europe, which use wood for cooking and heating. The reasons are complex and not always related to income. It can be an infrastructure problem. In Bulgaria, a gas distribution network was not sufficiently developed to enable universal access (Kisyov 2014). In Ireland rural households use peat for cooking and to heat their homes.[1] There are also high-income households who chose to use woodstoves for space heating since they see this as a green option (Bouzarovski and Petrova 2015).

[1] The 2016 Census reported that 5.3% of homes were using peat to heat their homes. There has been a slight increase of 0.5% from the 2011 Census (Eakins et al. 2022).

3.3 Energy Poverty

The concept of energy poverty was described in Sect. 2.3.1 within the context of the development discourse. However, its origins are actually within the Global North, when academic researchers in the 1990s introduced the concept as an aid to policymaking to illustrate the consequences of households having insufficient energy to meet their needs. This concept was originally called fuel poverty by researchers in the UK and Ireland (Boardman 1991).

A universal definition of energy poverty has proved elusive and contested. A major difference exists between the Global South and the Global North in the way energy poverty is understood in terms of its causes and the focus on the consequences of living in energy poverty. In the Global South, energy poverty is linked to poor infrastructure acting as a barrier to accessing modern energy sources, particularly in rural areas. Whereas in the Global North, while low-incomes are also considered a major factor in households living in energy poverty, the energy efficiency of houses is also considered to play a major role particularly in respect of a comfortable ambient temperature, since many of the buildings have poor insulation. Households living in energy poverty frequently use second-hand or old equipment with poor energy efficiency (Clancy and Roehr 2003). In the Global South low-income households in urban areas often live in informal settlements which are built of readily available, often scavenged, materials, such as wood and cardboard (CARE 1999). Not only are these materials flammable but they have poor thermal properties which influence the ambient indoor temperature.

A study in Japan, found that households experienced energy poverty at different times of the year linked to ambient temperatures and their geographical location (Castaño-Rosa and Okushima 2021). The north of the country has very cold winters and mild summers so energy poverty peaks there in winter whereas in the south of the country climate conditions are the reverse and energy poverty peaks in summer.

As long ago as 1976, Amory Lovins, the scientist and author of books related to energy choices, pointed out that people did not describe their needs in terms of electricity but rather more in terms of tangible items such as light, a cooked meal or a bus to take them to work (Lovins 1976 cited in Sovacool et al. 2012: 718), in other words, people describe energy services. However, energy policymakers when addressing energy poverty have a reductionist approach seeing the condition as consisting of only one or two dimensions which are of fuel or technological origin. In the Global South, energy poverty is not having an electricity connection or clean cooking. Whereas in the Global North, energy poverty is living in a home that does not have a comfortable ambient temperature. However, research shows that energy poverty is more complex and a multidimensional condition including gender and social inclusion aspects which are explored in the next section.

3.4 The Gender Dimensions of Energy Poverty

Access to sustainable energy can transform the lives of women and men in multiple ways contributing significantly to improving their quality of life. Access to sustainable energy can liberate men and women from drudgery and frees-up time for leisure, rest, and investing in improving knowledge and skills.

Energy interventions are likely to impact on women and men differently. As was pointed out in Chapter 2, women and men have different roles, responsibilities and influence within the household, markets and their community (World Bank 2005). This leads to differences in their access and use of energy, and the impact of energy services on their lives. For example, availability of electric light during the evening hours may improve the quality of life for some household members, by allowing reading and entertainment, education and information from radios, televisions and, if available, the internet while for other members it may simply extend the working day.

The high up-front costs of acquisition and running energy services is a contributor to households remaining in energy poverty. This raises a concern for female-headed households, which are often over-represented in low-income quintiles (World Bank and IEA 2015). However, one needs to be cautious about the universality of assumptions. The data presented in the 2015 Global Tracking Framework showed that, at the time of the survey, in 13 countries female-headed households had better access than male-headed households to electricity. However, no explanation was offered for this variation. In general, the difference in access between male- and female-headed households is smaller for non-solid fuels than for electrification.

In most parts of the South women compared to men suffer more severely from energy deficits and energy poverty (UNIDO/UN Women 2013). Women's capacity to control processes and resource allocation on many issues including energy is limited. Men and women often use, are impacted by, or benefit from energy services differently. The same energy service may have different social or economic outcomes for men and for women. For example, men may choose to locate a light outside the house for security reasons (such as protecting livestock from theft) while women may choose to locate the light in the kitchen (Cecelski 2000). Women and men have different perceptions about the benefits of energy. For example, research in Sri Lanka on the gender-related impacts of micro-hydro found that women saw electricity as providing the means for reducing their workload, improving health and reducing expenditure, whereas men valued electricity in terms of leisure, quality of life and education for their children (Dhanapala 1995 quoted in Barnett 2000).

Men make many of the decisions about buying household energy services which can have implications for household budgets. In South Africa, a survey found that some households would buy recreational equipment, such as TVs and radios before investing in labour-saving equipment for housework (Makan 1995). Similar findings are reported in Zanzibar (Winther 2012). Expensive batteries were bought for young men to listen to taped music (Makan 1995). Although batteries are convenient, they are a very expensive way of buying electricity, as in this form it costs about per $200 kWh (based on the cost of small batteries in the UK in 2023).

Most of the initial research related to energy poverty has been carried out in Europe, although studies in several Asian countries, including Japan, have been published (Castaño-Rosa and Okushima 2021). The social dimension of energy in Europe has focused on vulnerable consumers using the concept of energy poverty rather than energy access as formulated within the SDGs. This formulation reflects a perspective which considers that problems related to energy access in industrialised countries are related to capacity to pay for energy rather than with supply-side issues related to availability which are major factors in many countries of the South. There are studies, primarily qualitative, which have looked at differences in energy poverty levels across a range of social characteristics. Overall, the conclusion is that women in the Global North suffer from energy poverty more than men with the difference increasing with age (EIGE 2020). Self-reporting data from France about whether the respondent felt cold at home, as an indicator of energy poverty, found distinct gender differences. Those most likely to be living in energy poverty were single-parent families and people living alone (often in older age groups), predominantly women and living in rented accommodation (ONPE 2016). Funding of retirement appears to be a significant factor. Data from Spain show that prior to the age of retirement there is little difference between women and men's expenditure levels, however, this appears to change when retirement age is reached, for woman-headed households relying solely on pensions their annual expenditure on energy is between 10 and 15% of their income which reflects an increase in their energy expenditure as a percentage of their income compared to pre-retirement. Research in Germany highlighted age differences towards energy use. Elderly women were found to consume less energy than younger women (EPSECC 1997; Preisendoerfer 1999). Approaches to energy conservation are also found to show age differences; it is elderly women who change their behaviour patterns, for example, cooking less, to save energy whereas young women tend to opt for technological options.

3.5 Health

3.5.1 Health Impacts of Energy Poverty

There is a substantial body of research evidence about the health effects of household air pollution (HAP) produced primarily by inefficient stoves and their fuels. The World Health Organisation (WHO) estimates that HAP causes annually 4.3 million premature deaths worldwide (WHO 2014). HAP is the cause of more deaths than HIV/AIDS, malaria, tuberculosis and malnutrition combined (Lim et al. 2010). Women and children bear the heaviest burden due to their high exposure levels because of the time they spend in the kitchen. There are a range of medical conditions linked to HAP including cardiovascular disease, low birth-weight and perinatal mortality, eye diseases including cataracts and blindness, asthma, nasopharyngeal and lung cancers as well as increased risks of maternal depression (Putti et al. 2015).

Of the global two million annual IAP-related deaths from chronic obstructive pulmonary disease (COPD) and lung cancer, 60% of adult deaths are women (World Bank 2012). There is evidence that men's health can also be affected by exposure to air pollution when they spend time at home in the kitchen, a centre of cultural significance for socialising and eating together, or in the course of their work, for example, as chefs. Men's mortality risk increases when HAP-related illnesses combine with other health issues (World Bank 2012).

While HAP and cooking is a recognised health issue in the Global South, it is less well recognised in the Global North. However, in Eastern Europe, women are cooking and, during the cold winters, heating their homes with wood possibly with poor ventilation (Clancy et al. 2017). This is not a life-style choice but a situation of back-switching from subsidised state-supplied energy to market-based supply after the Eastern European departed from a communist system. Parts of Europe, such as Bosnia-Herzegovina, Kosovo and Chechnya, have experienced violent conflict with an associated breakdown of services including energy which has meant women having to cook with fuelwood (Clancy and Roehr 2003). The lack of awareness of the links between HAP and indoor burning of wood may mean that related illnesses are going unreported with a consequence that mortality linked to HAP is underestimated. Indeed, a survey in Ireland related to household solid fuel use found that the stated levels of knowledge of the health risks were low, especially among the primary users (Eakins et al. 2022).

It is not only policymakers who are unaware of the dangers of HAP, but also potential sufferers in communities most at risk exhibiting low awareness of the dangers of HAP (Bruce et al. 2006). A detailed study in two South African villages found that while women acknowledged the immediate effects of smoke, such as headaches, coughing and sore eyes, the long-term health impacts and the links with morbidity were rarely recognised (Matinga 2010). Villagers have a different conceptualisation of a symptom such as coughing from woodsmoke, it is normalised as part of everyday life. Possibly what is more of a concern, since they are in a position to create awareness of the dangers of HAP, is that health service professionals gave similar responses to villagers about the dangers of wood smoke (Matinga 2010). Matinga found, in her South Africa village study, issues related to firewood collection and HAP were not included in any of the health programmes and activities in the region where the villages were located. In part, this can be attributed to the training of nurses and programme managers of health services which is rooted in western conceptualisations of illness and prioritise health issues from this perspective. It is a positive that efforts are now underway to address the need for the incorporation of the health effects of HAP into medical research, curricula and clinical training programmes. The Global

Alliance on Clean Cookstoves (GACC) has started a training programme on HAP for healthcare professionals (GACC 2016).

There are a range of other health issues related to cooking and the associated fuels which receive less attention from researchers. Children spend significant amounts of time in the kitchen, leaving them prone to accidents related to open fires and stoves such as burns and scalds (Bruce et al. 2006; Matinga 2010). In South Africa, burns were found to be the leading external mortality cause for children under the age of one (Butchart 2000 cited in Matinga 2010). Paraffin also poses a danger to children who are at risk from poisoning through accidental ingestion, for example, in South Africa, in 2000, an estimated 4000 children died from paraffin poisoning (Butchart 2000 cited in Matinga 2010). The evidence from South Africa indicates that in the under-five age group, more boys suffer from burns and paraffin poisoning than girls, while in the 10–14 age group it is predominantly girls who suffer from burns. However, the risks for children will be context-specific, since culture will determine whether boys or girls will spend more time in the kitchen. Women also suffer from burns when cooking and men when they are drunk.

Fuelwood collection also has a range of health issues which receive little attention. The household gender division of labour generally allocates woman the responsibility for fuelwood provision which can mean, often daily, carrying loads of 20 kg or more over several kilometres. Fuelwood collection is an activity that women will perform for most of their lives, often starting as young as 6 and may continue until they are well over 70 years of age (Wilson 1961; Matinga 2008). The weights of wood carried well exceed recommended guidelines.[2] One of the few qualitative studies, based on a survey of 276 women firewood carriers in Addis Ababa, Ethiopia, found that women carry an average of 36 kg (about 75% of body weight) over 12 km (Haile 1989). At least 17% of the women were carrying loads heavier than their body weight. Girls, and sometimes boys, help their mothers which can result in their missing school, thereby damaging their future livelihood choices (Agrawal 1997; Vogt 2007). The

[2] The UK Government recommends that for adult men the maximum weight a man should lift is 25 kg when the load is held close to the body at around waist height. Above shoulder height the load should be at least 5 kg less. For women the recommendation is 16 kg for a load held at waist height. These figures assume that the load is easy to hold, preferably with both hands, also that the weight is being lifted by a person in good health and capable of physical exercise. https://www.beckettandco.co.uk/manual-handling-faq-weight/ (accessed 3 March 2022).

health consequences over their lifetime include back pains, eye and chest problems and high rates of miscarriage, as well as frequent falls, bone fractures, headaches, rheumatism, anaemia, skin irritations, poisoning, allergic reactions, fungal infections, snake bites and internal disorders (Haile 1989; Haile 1991 cited in Matinga 2010; Wickramasinghe 2003). Although this does not detract from the seriousness of this list of health issues, it should be kept in mind that it is based on qualitative research using participant recall and makes no attempt to offer rival explanations particularly for impacts such as backache, which could be due to other factors such as stress, poor posture, carrying other household supplies such as water. One of the few clinical studies of wood head-loading did find evidence of a higher incidence of prolapsed disc, but no other statistically significant degenerative skeletal changes, among bearers of heavy loads compared to a control group (Echarri and Forriol 2002 cited in Matinga 2010: 31). Chopping wood also produces a range of accidents of varying degrees of severity to hands, feet and eyes caused by splinters from which left untreated can develop into more severe complications (Matinga and Clancy 2020).

When men collect fuelwood, they often use animal or mechanised transport so are less prone to the physical injuries suffered by women. However, in urban fuelwood markets, men carry heavy loads so it can be anticipated that they are at risk from physical injury. Another potential health issue for single men in urban areas relates to eating in informal road-side restaurants which can be because of not having a partner to prepare meals for them. The food is not prepared under hygienic conditions, food may be stored incorrectly or reheated insufficiently. Their informal nature will probably mean restricted access to clean water and waste disposal (Clancy et al. 2006). In the Philippines, health official reported that seven out of ten food handlers were found to be infected with parasites (Lumampao et al. 2006).

Other health issues with which receive little attention in terms of their gender and energy connections include the impacts of energy poverty on people living with HIV/AIDS, malaria and other major diseases such as tuberculosis (TB). These diseases can reduce the capacity to undertake many of the household's basic tasks, such as wood collection, which require physical effort (ENERGIA 2006). In the rural South Africa women living with HIV/AIDs explained that, despite their concern about the effect of wood smoke on TB, their household responsibilities left no choice but to stay in smoky kitchens whereas men had the option to go elsewhere (Matinga 2010).

A second less recognised health issue related to energy poverty is the impact on mental health. Research in the Global North shows the people living in energy poverty are statistically more likely to report poor health and emotional well-being than people not living in energy poverty (Thomson et al. 2017). Living in energy poverty is associated with poor housing conditions which has an impact on the energy demand to regulate the ambient temperature which can result in a range of mental health issues, such as anxiety, stress and depression, arising from the tensions of trying to balance household bills and avoid debt (Snell et al. 2015). Based on self-reported data, a public health survey in Barcelona, Spain, found that women considered to be living in energy poverty reported poor mental health 1.9 times more often than those women who are not living in energy poverty. While for men poor mental health was cited 2.1 times more frequently by men living in energy poverty than by men not living in energy poverty (Novoa and Bottazzi 2018 cited in Birgi et al. 2021). There is anecdotal evidence that energy poverty can lead to feelings of isolation and loneliness because people feel ashamed to invite people into their homes because of the uncomfortable ambient temperatures.[3] Even healthy household members suffer additional stress when having to care for the sick, who may require more warmth, specific types of nutritious meals and more boiled water than healthy family members.

One of the health issues with an energy dimension related to households is the effect of ambient temperatures on human physiology and its functioning. The human body has self-regulatory mechanisms which allow people to function, with varying degrees of efficiency, across a range of temperatures. However, if the range is exceeded, there comes a point when the ambient temperature is too low or too high for the self-regulatory functions to respond effectively including impacting on the body's physiological functions and people are said to be experiencing cold or heat stress. Heating and cooling systems play a role in maintaining indoor air ambient temperature which in turn is influenced by the energy efficiency of the building. There is a concern that people living in energy poverty will struggle to maintain an ambient temperature that meets their requirements. Women are considered to have a greater sensitivity

[3] Http://www.coldathome.today/ (accessed 4 March 2022).

to ambient temperature than men (Brida undated). Young children and older people are particularly vulnerable to heat and cold stress partly due to physiology while for older people a lack of mobility and their behaviour (for example, reducing their food intake) are additional compounding factors (Chard and Walker 2016). A study conducted in 2014 in eight EU states bordering the Mediterranean found that 30% of the population reported that they were unable to keep their homes adequately cool in summer—70% of the cohort were above 65 years of age (Bouzarovski and Tirado Herrero 2014). There is also a gender dimension, at least in the EU, where more women heads of household live in older, less energy efficient homes than male heads of household (Elnakat and Gomez 2015). Gender is also a factor in deaths related to heat waves. In France, during a three-week period in August 2014 when day-time temperatures were above 40 °C in many parts of the country, 15,000 excess deaths[4] were reported. Disaggregation of the data showed that the excess mortality started at 35 years of age and gender differences began at age 45 when the number of excess deaths for women was 15% higher than for men of comparable age (Fouillet et al. 2006).

An issue which is known to exist, but for which there is little statistical data due to the nature of the issue, relates to women experiencing sexual and other forms of violence when collecting fuelwood and carrying out their daily survival tasks, including water collection and staple food preparation (Haile 1989; Potgieter et al. 2006; Matinga 2010; Porter et al. 2013). The sensitive nature of these issues makes it difficult to have any measure of the extent of harassment. Most of the available documented evidence comes from displaced person camps (MSF 2005; Kasirye et al. 2009). The International NGO Médecins sans Frontières (MSF) reports that in West and South Darfur, Sudan, between October 2004 and mid-February 2005, they treated nearly 500 women and girls who had been raped. It is estimated that 82% of the rapes occurred outside the camps while the victims were searching for firewood, water or animal fodder (MSF 2005). Cultural norms can discourage women from reporting sexual assaults or because firewood collection outside camps can be illegal in countries where camps are located (Lahn and Grafham 2015). During the Bosnian War in the 1990s, women in Sarajevo, while gathering fuel, reported incidents of physical and psychological violence,

[4] 'Excess deaths' is a term used which expresses a percentage of additional deaths in a given time period compared to a baseline period.

for example, from sniper fire (ESG and CSD-NGO WC 2001). Men can also suffer physical violence when trying to collect wood in such situations. A study in displaced person camps in Northern Uganda reported that men and boys were concerned about the safety of their female relatives so had taken over responsibility for fuelwood collection. They were also subject to physical assaults from armed gangs (Kasirye et al. 2009).

3.6 Energy Access as a Health Enabler

Access to energy in healthcare facilities is a critical enabler for well-being and can improve maternal care and facilitate childbirth deliveries. Every day, approximately 800 women die from preventable causes related to pregnancy and childbirth. Access to electricity in health facilities can increase the number of successful childbirth deliveries, especially at night. Electricity is needed for adequate lighting, but also for sterilisation and obstetric equipment. Unfortunately, healthcare facilities in the Global South experience similar problems with access to electricity as households. World Health Organisation (WHO) data for 2023 gives approximately 12% of healthcare facilities in low- and lower-middle-income countries of South Asia as having no access to electricity while for Sub-Saharan Africa it is around 15% (cited in World Bank 2023: 53). Again, as with household access, rural areas suffer more than urban areas, although even in the latter access, particularly to a stable supply, can be problematic.

Nutrition is a key component of human health. There are other technologies beyond stoves which can play a role in promoting healthy diets. In rural Tanzania, children's consumption of foods rich in vitamin A[5] was reported to be higher in the families who adopted a solar dryer than in those who had not adopted it (Mulokozi et al. 2000 cited in Blackden and Wodon 2006: 24). The portable solar dryer for vegetables was introduced with the aim of improving food quality in terms of reducing contamination and increasing the vegetables' nutritional content compared to the

[5] According to the WHO, Vitamin-A deficiency is the leading cause of preventable blindness in children and increases the risk of disease and death from severe infections. Vitamin A deficiency is found in more than half of all countries, especially in Africa and Southeast Asia, with young children and pregnant women in low-income countries particularly at risk. https://web.archive.org/web/20190816130048/https://www.who.int/nutrition/topics/vad/en/ (accessed 4 March 2022).

existing method of open-air drying. A key factor in the dryer's adoption was involving the women in its design including their criteria for acceptance which were flexibility, portability, time-saving and individual use.

There is some evidence to suggest that access to electricity has an impact on family size (Grimm et al. 2015). Data analysis aiming to correlate family size and access to modern energy found a stronger link with electricity than modern stoves (Belmin et al. 2022). However, the mechanisms influencing decisions about family size varies. Electricity powers modern communication systems, such as the mobile phone and television, which can help disseminate information about family planning and, particularly TV, can help show positive images of families with small numbers of children. The influence of information and diffusion of new norms cannot be taken as a given since other factors can come into play. For example, being able to determine family size depends on a woman having access to contraception and that she has the agency to use it. Being able to participate in labour markets could be the driver for a woman to control her fertility. On the other hand, the input of child labour into meeting family needs could result in no attempts to limit family size. Whereas the direct and indirect costs of children might have the opposite effect.

3.7 Time Poverty and Drudgery

Time poverty can be defined as the condition in which an individual does not have enough time for rest and leisure allowing the body to recuperate with negative implications for health. This is an issue that attracts a lot of interest in the development discourse. Firstly, increased time for the body to rest and to experience reduced physical effort brings improvements in well-being.[6] Secondly, freeing up time provides increased opportunities for income generation which within the gender and energy discourse is associated with women's empowerment. Free time can be used for working towards political empowerment by attending adult education classes or school and engaging in civic activities (United Nations Millennium Development Project 2005).

[6] There is no universally agreed definition of well-being. Here I am using a broad definition in which a range of conditions, such as freedom from violence, and assets, such as education, contribute to good health and quality of life.

Most of the available evidence related to time poverty relates to women with little available about men. Data indicate that a working day for women between 11 and 14 hours is not uncommon, while for men it is around 10 hours (ENERGIA 2006). Figures are usually based on estimates of respondents' recall which is notoriously unreliable. Women's time poverty is related to fulfilling their household tasks, particularly fuelwood collection, fetching water and food processing. Fuelwood collection can be time-consuming. For example, data for Zambia indicates more than 800 hours a year, while for Gambia and Tanzania it is around 300 hours (Malmberg Calvo 1994 quoted in Rossi and Lambrou 2008: 10). Times vary depending upon location, season and terrain. While access to modern energy sources might seem a gain for women in terms of time saved and reduced drudgery, it can add burdens in other ways. In India, women report that the trip to the LPG dealer results in the loss of the husband's daily wage. The response is that they use the LPG sparingly, supplementing it with biomass fuels (MSSRF and CRT Nepal 2019).

It appears that the situation in respect of the burden of care work is no different in the Global North. Within the EU, women are responsible for 75% of unpaid household tasks (European Commission 2020). While women maintain the major responsibility for childcare, as well as providing support to older relatives and those with disabilities, the COVID-19 pandemic did see men contributing more time to family care (EIGE 2022). The nature of this work has an energy cost, both within the home and outside for accessing services (as discussed in Sect. 3.4). Many household tasks, such as washing clothes and preparing family meals, are mechanised with an associated energy demand which can require decisions about how often and when (especially if the household has a variable day/night tariff) to perform these tasks. Since it is mainly women who perform these tasks by default, they become the family's energy manager. When this role leads to requests for behavioural changes in response to energy price increases, such as switching off lights and entertainment equipment, it can lead to tensions within the family (Petrova 2017).

Water collection may be more time-consuming than fuelwood collection. A study in the early 1990s in the resource-deficit Chiduku Communal Area in eastern Zimbabwe found that women spent 4.1 hours a week on collecting fuelwood and 10.3 hours on water. In meeting these basic household needs women provided 91% of the household's total effort (Mehretu and Mutambira 1992). Similar findings are reported in both urban and rural areas of Benin, Ghana and Madagascar (Charmes

2006). Although piped water is available in urban areas, it can require queuing at public standpipes which can be time-consuming. In Manila, Philippines, to meet their daily water needs women in focus group interviews report that they often had to start queuing before dawn (Approtech Asia 2005).

It is not only the time allocation but also the associated drudgery (as well as health issues described above) for women's household tasks which are mainly fulfilled through their own physical labour. Again, women seem to outperform men. Village transport surveys in Ghana, Tanzania and Zambia found women carrying on their heads about four times as much in volume as men, consisting primarily of water, firewood and crops destined for grinding (Blackden and Wodon 2006). The daily task of grinding grain to prepare staple foods is a manual task mainly done by women. In rural Mali, a woman can spend between one- and three-hours daily grinding grain or they can wait in line at the grain mill (Porcaro and Takada 2005). A study in Northern Tanzania found that converting grain mills from running on a diesel engine to electricity saved women's time queuing for milling. Although the amount of time saved was not quantified, indications of the level of the significance of those savings comes from evidence that women now had time to set up their own small enterprises (Maleko 2006).

However, the evidence shows a more complex picture of time and drudgery related to context. As was shown above men and children can also be involved in household fuelwood collection. Evidence from Ethiopia, Madagascar, India, Nepal, Vietnam and Indonesia shows a mixed picture, sometimes it is both men and women collecting fuelwood, while sometimes the primary collectors are men and children (Cooke et al. 2008). Responsibilities shift when the fuelwood supply close to the household decreases, then men tend to take over (Cooke et al. 2008). During labour-intensive periods of the agricultural cycle, such as harvest time, children can be allocated fuelwood and water collection duties which if this keeps them out of school potentially damages their future livelihood choices (Agrawal 1997; Blackden and Wodon 2006; Vogt 2007). In urban areas, such as in Benin and Ghana, men were found to provide household fuels (Blackden and Wodon 2006). As one group of researchers concluded the household decides 'who will collect and at what time in a manner that minimizes the cost to the household' (Cooke et al. 2008: 113).

Evidence from rural Ireland, shows that in households using peat or turf, it is often a man's responsibility to provide the fuel as well as lighting, maintaining and cleaning out the fire (Eakins et al. 2022). For older men, these tasks represent the traditional male role of provider for the family, as well as a link with the past when turf cutting would have been a daily activity. For younger men working in offices, the physical activity acts as a stress relief.

Two technological interventions have been considered to address the issues of time poverty and drudgery: improved cooking and electric light. However, it is unclear (again because of the reliance on re-call data) as to how much time a new stove saves. Two studies, which used observation methods for data collection, when LPG replaced kerosene, reported 8 minutes (Barnes and Sen 2004) and 15 minutes (Saptyani 2010). Savings are probably not as large as predicted since households continue to use multiple fuels and do not abandon fuelwood entirely.

A survey in a poor urban township in South Africa, found that 216 of the 250 households surveyed (86%) had an electric stove or hot plate, although no data were given on frequency of use. Women reported that an advantage of cooking with electricity was that it is faster than with kerosene (Annecke undated).

Good-quality electric light allows women greater flexibility in time management, for example, tasks can be carrying out simultaneously and completing a task when it best fits within the day. Understandably there are concerns that this does not appear to be changing time poverty. However, women, the targeted beneficiaries, do report that they value this flexibility (Barkat et al. 2002; Annecke undated; HDRC 2002; Winther 2008). One woman reported that having a solar lantern for cooking meant she did not have to use the torchlight from her mobile phone which was now free to provide music or make phone calls (Soria et al. 2016).

Women's time saved by using electric light comes from no longer having to go to buy kerosene as well as cleaning and maintaining lamps (Mahat 2004; Standal 2010). Time-saving for women through the introduction of other fuels occurs in other ways. The availability of LPG in rural India has brought a shift in gender roles. While this has saved women time in fuel collection, if there is no household delivery of LPG cylinders, the man of the household goes to the dealer to collect the cylinder (MSSRF and CRT Nepal 2019). However, switching fuels may not always reduce overall time poverty. With the introduction of biogas digesters women report reduced time in fuel collection. However, each kilogram of dung needs mixing with approximately one litre of water which has to be brought to the digester. It is usually the task of women to carry the water from the source to the digester which results in little or no overall time-saving compared to fuelwood collection (Clancy et al. 2012).

In one of the few studies which also looked at the changes in men's time poverty relative to women's, found that in China, it was men who benefitted from electrification of grinding and milling, which reduced women's time used for pig food preparation. Women's freed-up time could then be used for men's agricultural work which men saw as an opportunity to look for better-paid work in urban areas. Consequently, when women took over men's work their overall workloads remained about the same as prior to electrification (IDS 2003). On the other hand, in some parts of Asia where there is paid labour in agriculture, for example Indonesia and Bangladesh, evidence suggests that mechanisation had a negative effect on women's employed labour. In Bangladesh, when manual rice-hulling was replaced by steel rice hullers many rural women lost their jobs that contributed vital household income. Landless families were hardest-hit since women's work in rice-hulling was estimated to provide 55% of their annual income (Cecelski 2004).

Community services can also contribute to saving women's time and labour. In rural Nepal, installing a hydro-powered grain mill resulted in women no longer having to rise at 4 am and spend at least two hours processing grain, although they do have to travel to the mill and wait to be served which is time that can be passed socially (Mahat 2004 cited in Dutta 2005: 14). Piped water supply can significantly reduce time spent on fetching water. In Zanzibar households reported saving between 20 and 25 hrs per week when electric water pumps and taps were introduced into the village (Winther 2008).

3.8 Promoting Energy Access and Addressing Energy Poverty

The two obvious factors which influence energy access are availability and affordability. However, a third factor is quality of supply, which includes reliability and convenience as evaluation criteria, is also found to be a determining reason for choosing a particular energy source or supplier.

The affordability of electricity is governed by initial connection charges, cost of wiring and the standing charges. Researchers found significantly large differences in connection charges between countries

(University of Oslo et al. 2019). For example, in 2016, in Kenya it was US$350, while in Nepal it was US$91. In Kenya, the connection charge was found to be a particular access barrier for widows and single/divorced women.

Subsidies are a common policy instrument to help with energy costs. However, they do not always reach the target group, getting lost along the supply chain. For example, in Bangladesh, households were paying 14% more than the official subsidised price for kerosene, a situation that was linked to poor enforcement by the authorities (IISD et al. 2019).

Utilities are often reluctant to provide a service where there are doubts about the legal tenure of property and where the dwelling is not considered to be a permanent construction. Many low-income households fall into one or both categories. The consequence can be getting a connection illegally or through a third party, often a 'slum landlord or lady' who has a legal connection and charges exorbitant rates to tenants for their supply (Annecke and Endelli 2006). Not all electricity used through illegal connections is with the explicit compliance of the end-user. Research in Ghana found poor urban households were the victims of deception with unscrupulous fellow residents making illegal connections but collecting the payments on the pretence of making the payment to the utility (Bannister 2002). Illegal connections are a problem for both utilities and households. For the utilities, illegal connections mean unpredictable loads, damage to the infrastructure and a loss of revenue. For example, in Bahia State, Brazil, it was estimated that, at the beginning of this century, around 11% of the electricity distributed was diverted to illegal connections (Andrade 2004). For the consumer, illegal connections are unreliable and dangerous with connections being made without the usual safety mechanisms.

3.9 What Changes After Improving Energy Access

There are two points to keep in mind when assessing the changes that occur as the outcome of an intervention such as improving energy access. Firstly, change can take time to appear. For example, a study of the influence of electrification on women's employment in the US between 1930 and 1960, showed that while the number of girls going to school increased because of electrification, it was the second generation that reaped the benefits in terms of employment (Lewis 2014). Secondly, much of the available evidence is based on recall which can be unreliable.

Evidence shows that all household members benefit from improved access to services delivered by modern energy sources or technologies, with the nature of the benefits varying with gender and age (Cecelski 2004, 2005; ENERGIA 2006, 2019). The evidence has tended to focus on the changes to women's (and sometimes girls') lives. A rare study gives some insights into the changes in men's lives, in Zanzibar, before electrification, men spent their evening hours with their male friends outside the home. However, after households had access to electricity, men were more inclined to spend their evenings watching TV at home, although sometimes this would be at a friend's house (Winther 2008).

The time saved thanks to improved access to energy is often used differently by men and women and not necessarily in ways interventions intend. In general, it appears that men are more likely than women to use these savings for recreation and leisure. Women tend to use the time for housework and childcare, as well as for resting, socialising and watching TV, and not necessarily for income-generating activities (Matly 2003; Clancy et al. 2012; Fernández-Baldor et al. 2015). A study in Namibia found that women stayed up later than men, not working but socialising (Wamukonya and Davis 1999).

Whether women meet the expectation voiced by many involved in development practice, that they will use their time freed through energy-saving devices for income generation is not a given. There are some women who do so willingly, as we see in the next chapter, even though this can be linked to a longer working day. Participating in income generation is not necessarily seen as an exercise of agency but one of a need for cash (Meikle 2004). In Sri Lanka it was found income-generating activities were not a high priority for women. Instead, they used the two hours electricity gave them for doing better housework and childcare, as well as resting, socialising and watching TV (Matly 2003). There was a similar reaction in South Africa, where women said that they were tired, harassed and wanted to rest and had no interest in taking up new activities (Green 2001). Evidence from the US suggests that women use the time saved when they acquire household appliances does not necessarily translate into them seeking employment but that the time is invested in their daughters who benefit in terms of their participation in labour markets (Lewis 2014).

There is some evidence from three econometric studies which examined the correlation between electricity in the home and women and men's formal employment. In two cases, there were positive increases in access to formal employment for women (South Africa [Dinkelman 2011] and Guatemala [Grogan and Sadanand 2013]) whereas the third (in India) only men benefitted (van de Walle et al. 2015).

Research from the US about the impact of time-saving through electricity access indicates that acquisition of labour-saving devices lowered the opportunity cost of health-promoting activities such as sterilisation of nappies (Lewis 2014).

The process of gaining access to modern energy sources can itself be transformative and contribute to women's empowerment. A project in Bihar to promote access to better quality lighting through Solar home systems (SHS),[7] would only allow women members of self-help groups (SHG) to purchase the systems (University of Oslo et al. 2019). Participation in the project has enabled women to learn to sign their names, to speak out during group meetings about their problems and to manage money. This has helped challenge gender norms. Men see women as more confident and independent which has enabled women to be active outside the home without a male chaperone.

Do households make the transition to modern stoves?

As was pointed out in Chapter 2, the idea of a situation, visualised as a ladder, in which households moved from the bottom rung representing the most polluting and least efficient stoves through an intermediate stage, abandoning the old stove, to finally reach the top rung using clean, modern energy in high efficiency stoves has long been discredited. Households, across all income groups, do not completely abandon their 'old' technologies but continue to use them for a variety of reasons (Masera et al. 2000). This retention of the original technology also includes kerosene after switching to electricity (van de Walle et al. 2015). A study of an electrified rural village in the Northwest province of South Africa found that, on average, 3.6 years after connection to the grid, 44% of the electrified homes had never used an electric cooker (Rollin et al. 2004).

[7] Lighting a Billion Lives https://www.gogla.org/lighting-a-billion-lives (accessed 1 June 2022).

It tends to be wealthier households who made the transition to electric cooking. A similar conclusion was reached about the transition from wood to LPG in several Asian countries[8] with distinct differences also between rural and urban areas where the transition is slower in the former, while even more restricted in remote rural areas (Palit et al. 2014 cited in MSSRF and CRT Nepal 2016). This situation is replicated across the countries of sub-Saharan Africa (see Sect. 3.2) (Ato Armah et al. 2019).

Independent evaluations of stoves programmes are not common particularly related to long-term use. A systematic review of the literature found that most of the studies identified focused on small-scale projects and programmes provided information based on a short time scale (Puzzolo et al. 2013). One study related to solar cookers, found that as soon as the project promoting them ended, households stopped using them regularly (GTZ 1999). Indeed, in some households the shiny reflective panels were found on bathroom walls as mirrors!

The factors which influence the adoption of energy technologies are complex and not entirely related to cost. Explanations related to income vary. The value of women's labour has been advanced as one argument for households continuing to use fuelwood (see, for example, Nathan and Kelkar 1997; Köhlin et al. 2011). While in high-income households, cooking may not be the responsibility of the senior woman in the family but is the task of a relative or hired help hence the cooking medium is not a priority issue. However, households do not always buy the energy technology themselves. Gifts from family members is one such route to access in which having a modern stove can be seen as a status symbol to show a family's progress (Annecke 1999; Masera et al. 2000; Winther 2008; Saptyani 2010). Nevertheless, the stove must match cultural practice related to cooking in terms of the quantities a stove can cook that allows the family to eat together. Cooking certain foods on wood or charcoal imparts a certain flavour which is not produced by LPG or electricity. Solar cookers also were found to conflict with cultural practices since they need to be operated at midday which does not coincide in many cultures with the family's main eating time of the evening (Mandhlazi 1999). Women also found it too hot to cook in the middle of the day (Annecke 1999).

[8] China, India, Nepal, Pakistan and Sri Lanka.

As was pointed out earlier, men make decisions about acquiring household goods and stoves are not an exception (Tucker 1999). Cultural norms can also play a role. A qualitative study in rural Bangladesh provided insights into decision-making about women securing an improved stove (Moniruzzaman and Day 2020). Women had tried to persuade their husbands to buy a stove using arguments about health, including saving on medical bills, but men were not convinced. Some argued that generations of their female relatives had used wood and dung without problems, a position the husbands' mothers supported. Culturally, in rural Bangladesh men would not argue with their parents, so would not contradict their mothers if they stated their daughters-in-law did not need an improved stove. Women reported that they had not asked for an improved stove to avoid confrontation.

Another example of income level not being the sole determinant of adoption of clean cooking technologies comes from the analysis of data for 31 Sub-Saharan African countries by Ato Armah and his colleagues which found that education level could also be influential. They found access to clean cooking was more likely for households that were wealthy <u>and</u> highly educated (secondary level or higher) compared with households that were either wealthy but with low education or poor but highly educated (Ato Armah et al. 2019). The implication of a low education level is that the household would be unlikely to be aware of the harmful effects of the combustion products of dirty fuels.

Whether female-headed households are more likely than male-headed households to adopt clean cooking is not clear. Some studies indicate a slight advantage for women-headed households (see, for example, Muller and Yan 2018), whereas others consider data not to be statistically significant (see, for example, Abebaw 2007) arguing that these households often are in the poorest quantiles so would be unlikely to have sufficient resources to buy the stove and fuel. The influence of the household head's age is also contested. While some researchers find that households in which the head is an older age group are more likely to adopt clean cooking than those with a younger head (see, for example, Ato Armah et al. 2019), other researchers say that older household heads may prefer to keep using fuelwood since it has strong cultural traditions (see for example, Rahuta et al. 2014). Similarly, there are mixed findings for household size. Some studies find that larger households are less likely than smaller households to use clean cooking fuels (see, for example, Ato Armah et al. 2019), whereas others find the opposite (see, for example,

Ouedraogo 2006). An explanation for larger households continuing to use charcoal, at least in the context of Mozambique, is that older members tend to do the cooking, who might prefer to remain with a fuel they are familiar with (Castán Broto et al. 2020). Charcoal may also be more appropriate for traditional dishes which require slow cooking.

Adoption of a stove is no guarantee of continued use. A significant factor leading to rejection of improved cook stoves is insufficient user training on stove (and chimney) use, cleaning and maintenance, which influences functionality, creating a frustrated user who abandons the stove (Puzzolo et al. 2013). For LPG stoves providing user training for safe use of the gas was found to reduce fears of explosions and positively influenced demand (Bates 2009).

What use is made of electric lighting? Certainly, lives are changed in numerous ways through alterations to time use with electric light enabling people to reschedule and reorganise everyday routines. Better-quality lighting which electricity provides compared to candles and kerosene can make household tasks safer (for example, not having to hold a kerosene lamp in one hand while trying to use a knife in the other hand). It also allows for multi-tasking (for example, doing a task in one part of the kitchen while watching pots on the stove in another part) which can lead to more efficient time use. Women in a study in rural India estimated that this new method saved them around one and half hours per day (Chakrabarti and Chakrabarti 2002).

There is evidence about electric lighting transforming the relationships between husbands and wives. First, it can reduce tensions within the household. For example, in rural Afghanistan, solar lanterns allowed for cooking outside, and the family could eat together in a smoke-free atmosphere and without the smell from kerosene lamps. Men also considered that the food tasted better (Standal 2010). Better illumination from electric light, allows people to see each other's faces which also convey meaning adding to the spoken word. Evening hours are the time for discussing family issues and good quality light enabled men and women to see their facial expressions which they felt brought better understanding about what was being said, so reducing tensions (Standal 2010).

Electric light also influences opportunities for intimacy. It can increase, for example, when more than one room is lit, parents in Afghanistan reported that they could retire to a separate room from their children (Standal 2010). However, there was the opposite effect in Zanzibar, where opportunities for intimacy decreased if men stayed up late either at home entertaining guests and watching TV or going to a friend's place, while their wives went to bed (Winther 2008).

Tensions within households for access to light are reduced with electric light since the illuminated space is larger than with candles or kerosene so conflicts are avoided about who will have access to light since several people can use the same light at the same time (Winther 2014).

There are two often encountered claims in the advocacy literature about the influence of electric light on women's lives for which the evidence is limited and at best shows correlation not causality. The first claim relates to street lighting and the second claim relates to women's literacy. In relation to street lighting, the assumption is that women will feel safe and therefore go out at night (see, for example, ENERGIA 2006; SEforAll 2018). The evidence is ambivalent. Women in Afghanistan report that solar lanterns allowed them to visit other households for celebrations (Standal 2018).

People report that with streetlights in place they feel safer from a range of threats such as thieves, snakes and other dangerous animals, and evil spirits (Winther 2008; Standal 2018) which does not translate automatically in leaving the house for walking in the streets since cultural norms, particularly as relate to notions about women's morality, can act as a barrier. For example, the social norms of the hill tribes in Northern India about women who go out after dark did not change when street lighting arrived in their villages. These norms continued to act as a barrier to women leaving their home in the evenings (Kelkar and Nathan 2007).

In respect of the second claim linking women's literacy to electric light, again the evidence is not overwhelming. The claim is often based on other things being in place, for example, that street lighting will enable women to go out at night when they have completed their chores to attend evening classes (see, for example, Davis 1998). Such a notion also assumes that someone is providing these classes (Annecke 1999). What is taught is often practical, such as 'mothering skills' and hygiene to improve family health (Standal 2018). What women read may not contribute to consciousness raising, for example, in South Africa, the Bible, magazines and newspapers (Annecke undated). Although what men read seems not to have been an issue of interest. When it comes to children doing homework by electric light, studies in South Africa found contradictions between surveys where respondents state that it is the case whereas participant observation found little evidence to support this assertion (Annecke, 2005; Matinga, 2010).

There are also claims about electric light and children's access to education and their subsequence performance. A study in Tunisia, found a link between increased school attendance by girls linked to street lighting. Families reported that they felt girls could walk more safely to school in the early morning which was also supported by school administrators (Cecelski et al. 2005). A study in the US found a correlation between school attendance by teenage girls and a household's access to electricity (but not specifically electric light) (Lewis 2014). However, in Sudan, girls could not go to night classes due to concerns about girls' safety in the dark (Ahmed 2008). Although schools can be creative in overcoming these barriers. In Zanzibar, in the month before important exams, boys and girls could attend night classes and sleep in gender-segregated dormitories with teachers (male and female) supervising them (Winther 2012). However, it can be difficult to demonstrate causality, due to the existence of other confounding factors. Increasing girls' school attendance requires a reduction in the level of household chores (or a reallocation of tasks—see below) they are assigned which tend to be much lower for boys. For example, the installation of electrified water pumps and taps in Zanzibar villages, freeing girls from their previous task of collecting water, has contributed to near equality in numbers in school attendance (in 2001, in Uroa village, 93% of girls between 7 and 17 and 95% of boys attended school [Winther 2008]). It might not be electric light but other pieces of equipment which increase school attendance, for example, in Bangladesh this was linked to the availability of fans providing a more amenable ambience for studying (Barkat et al. 2002).

Electric light at home can result in a long day for girls. They can go to school but have to do chores in the evening enabled by having better-quality light (Standal 2010; van de Walle et al. 2015). It appears that electric light at home has more influence on promoting girls school attendance than boys (Winther et al. 2016).

What influence do telecommunications have?

There is some evidence to suggest that when women and men spend together listening to the radio or watching TV strengthens women's position in the family (Ramani and Heijndermans 2003; Winther 2008). Men are more inclined to come home rather than spend time socialising with their friends. TV and radio are both sources of news, practical information and entertainment. Anecdotal evidence from South Africa points to the role of 'soaps' in providing inspiration for women to become financially independent and advice on dealing with relational problems (Anneke undated; Matinga 2010).

Programmes aimed at women impart information over a range of topics which they do not always put into practice (Winther 2008). Health promotion seems not to mention household air pollution (Clancy et al. 2012). One cannot assume that information alone will be sufficient to change social norms about fuelwood use (MSSRF and CRT Nepal 2016). The extent to which influence programmes promoting family planning have on family size is difficult to determine since data gathering about these matters is problematic due to their sensitive and private nature meaning that these are not readily discussed with strangers, particularly in rural areas. The limited evidence finds both positive influence and no influence depending on the survey location (IEG 2008; Clancy et al. 2012). There is a similar situation in respect of the evidence related to domestic violence (Standal 2010).

Parents the world over are concerned about the influence of TV on their children—on the one hand it has a positive influence as a source of education, on the other hand, not everything children watch might be considered beneficial. Nevertheless, parents report that TV plays a role in keeping children at home away from the dangers of the street, so it reduces their anxiety (Ahmed 2008; Matinga 2010; Annecke undated). Where there was no internet access, the radio had a positive impact on children's education during the COVID-19 pandemic when schools were closed, children could still have access to lessons (Behsudi 2021).

Having a television in the households brings a change in social practices. Adjusting meal and socialising times allow the family to watch favourite programmes on TV (Matinga 2010). Women can combine watching TV with other household chores, such as sewing, which they found relaxing (Winther 2008). The place of socialisation for the family also changes, from the kitchen to the room where the TV is placed. Interestingly, if the household continues to use its traditional stove, such a relocation might reduce exposure to HAP. In cultures where customary social principles for organisation include strict gender segregation inviting guests to see your new status symbol, that is, your television, creates a potential moral dilemma if this requires men and women guests to be in the same room. However, people in Zanzibar appear to have found a solution that satisfies these tensions. The room with the TV is divided according to sex with women sitting down one side and men along the opposite side. The room can often be seen from the street, which with electric light enables the community to see clearly that social norms are complied with (Winther 2008).

There is also a financial cost to running a TV. A study in a dozen countries showed that light and television constituted 80% of households' electricity consumption (IEG 2008).

Mobile phones allow women to keep in contact with family members particularly at important in times of need (e.g. food shortages) or to ask husbands working away from home for money to pay for school fees or social obligations such as weddings (Ahmed 2008). Internet banking via mobile phones has had a significant impact on access to money both to meet family daily needs and to enable businesses, particularly for women whose mobility is restricted by cultural norms (Winther et al. 2016). A study of Kenya's mobile payments system found that, between 2008 and 2014, the system helped lift 200,000 families out of poverty, which is approximately 2% of the country's households (Suri and Jack 2016). Nevertheless, there continue to be Gender Gaps[9] in mobile phone ownership in which the level of ownership is generally higher for men than women. As of 2019, there is a gender gap in Sub-Saharan Africa of 17% whereas in South Asia it is 27% with only slight improvements over a three-year period (GSMA 2020). Internet access has several barriers, with literacy and content in local languages being a primary cause. The issue of literacy reminds us that women are not a homogeneous group. Among women it tends to be the young, well-educated woman, in a profession or owns her own business and lives in an urban area (where the electricity supply is more reliable) who uses the internet (GSMA 2010).

There are also variations in the Global North, at least within the European Union, in mobile phone and internet use. The urban and suburban areas of 22 Member States had better access than the rural areas (Eurostat 2021). A study in Lisbon, Portugal, found that users of mobile phones and the internet tended to be men, younger than 75 years of age, with higher education currently, or recently, employed with a good income (Barbosa Neves et al. 2013). Explanations by older respondents for why they did not use a mobile phone and the internet were that they 'didn't

[9] The Global System for Mobile Communications Association (GSMA) calculates the gender gap using a formula: (men owning mobile phones as percentage of male population—women owning mobile phones as percentage of female population)/men owning mobile phones as percentage of male population (GSMA 2020: P7).

know how to use it' and/or 'felt that they didn't need it'. People in the older cohort who could use the internet had generally acquired the skill, familiarity and confidence to use computers as part of their work which could explain older women's lack of internet use. For older women, social norms when they were young would have influenced them to cease paid employment on marriage or on starting a family, so they probably stopped work before digital technology became widespread. Alternatively, they may have worked in jobs that did not use IT.

Access to social media through mobile phones can contribute to women's political empowerment by raising awareness of issues. In Kosovo, mobile phones helped mobilise rural women at short notice to contribute to developing a new constitution (GSMA 2010).

So what role is energy access playing in shifting gender roles and relations?

Within most societies, women have the primary responsibility for household tasks of caring for the family, feeding children, cleaning the house, cooking and washing clothes and dishes. Men's role is as breadwinner with the primary responsibility to provide food for the family. These roles become internalised; women consider themselves better at caring than men who in turn believe that caring is not something they could do. However, gendered roles change do over time as this quote from a woman in Ghana illustrates:

> I know a bank manager at Sunyani, who will wake up in the morning, wash his baby and wife's clothing while his wife stays in the kitchen preparing breakfast. When he is done, he dresses and goes to work. This makes it easier for both of them to go to work early. The men are supposed to help in household chores. (IDS and GIZ 2019)

There are changes in gender roles when there is a switch in cooking fuels which has been found to bring men into the kitchen particularly when it involves using electricity. This can also reduce women's stress related to food preparation, when men cook for themselves when their wives are busy (Kelkar and Nathan 2005). Men seem more reluctant to use other modern energy sources but there are reports from rural India that men cook with LPG and rice cookers in Nepal (University of Oslo et al. 2019).

Men will also help with other domestic chores when they can use electrical appliances, for example, ironing (Annecke 1999; Massé 2003).

Ironing provides an illustration of how the adoption of a new technology is influenced by whether its use conforms with gender norms of acceptable behaviour by which women and men are perceived, and judged, by others. In Zanzibar, notions of aesthetics govern perceptions of how a man should look when performing his tasks within his gender roles—he should be attractive (*hapendezi*). Ironing was considered an acceptable activity for a man since he was deemed to conform to these aesthetic norms (Clancy et al. 2012). Even so, electrification does not always lead to changes in the gender division of labour, for example in Bangladesh (Barkat et al. 2002) and Japan (Kelkar and Nathan 2005). In the US, the acquisition of new household appliances brought a reversal in the contribution men made to housework, reducing their input, while for certain tasks that had previously been a shared responsibility became women's sole concern (Hawthorne 1996).

Indeed, the desire for access to electricity is bringing a change in cultural practices in relation to marriage. In Zanzibar, cultural norms of what constitutes a 'good husband' in terms of what he should provide can now include an electricity connection. This expectation has resulted in some men delaying the age at which they marry since the cost of the electricity connection contributes to the costs of assets that men have to accumulate prior to marriage. Another cultural practice undergoing change is the number of wives a man will marry. In the past, a man's success (and hence status) was judged by his capacity to provide for several wives living in their own households. The cultural norms expect all wives to be treated equally, which would now include an electricity connection something that men are finding difficult to achieve. So, to avoid the risk of being ridiculed, men are choosing to marry fewer wives (Winther 2008).

To outsiders intent on producing improvements to living conditions through addressing energy poverty, cultural norms (which form part of gender roles and relations) can be overlooked and in doing so may form a barrier to adoption of technology or behavioural changes. Existing gender norms create identity and when a new technology is introduced that threatens that identity there is likely to be resistance to adoption of that technology, what Rogers calls a 'gender-tug-of-war' ensues (Rodgers 1994). When trying to introduce solar cookers in Zimbabwe, men are reported to have rejected the technology. One explanation is that men had reservations about how women would use the time saved in fuelwood collection by the introduction of the cookers (Green 2001). Another explanation is to see how gender identity can feel threatened and lead to uncertainty about gender relations. In Zimbabwe, traditionally technology and its development are seen as male preserves and part of their identity is to provide for the household (Nyoni 1993). So, when a project introduces an energy technology, bypassing men in their dissemination mechanisms, it can leave men uncertain about their new position with respect to women. Women also had their reservations about solar cookers since cooking would now be outside which removes women from the kitchen. In many societies the cooking fire is seen as the social hub of the family allowing the women of the household to socialise with their families. A shift to cooking outdoors can lead to a breakdown of the family's social web, thereby weakening women's influence within the family, hence why women rejected solar cookers (Green 2001). This contrasts with the experience of Afghan women using solar lanterns reported above.

For women, uncertainty comes when they feel their identity being threatened by the acquisition of new technologies such as an LPG or electric stove which no longer requires wood collection. In any culture an artefact, such as a stove, is used to assess situations as well as to make judgements about whether an individual is conforming to the prevailing norms and values. In her ethnographic study of gender and energy in a Xhosa village, Matinga found that an improved stove could be used to judge whether a woman is a 'good woman', implying that by not using wood she is lazy or too modern (Matinga 2012).

TV and radio are reported to have impacts on gender identity. The positive images and stories about women, including an understanding about women's legal rights, have changed women's (and to some extent men's) perceptions of gender roles and relations (Annecke undated; Chaieb and Ounalli 2001; Standal 2010). Women learn that they 'don't have to remain as second-class citizens' (Barnett 2000) which provides them with the confidence to challenge male authority (Standal 2010), although it can be resisted by some men who resent their status being undermined (Annecke undated).

Whether this greater awareness of women's rights has led to a decline in domestic violence is however difficult to assess due to norms which determine which matters it is acceptable for discussion with outsiders. A study of remote Afghan villages, based on anecdotal evidence that the relationships between men and women had improved due to TV, considered that domestic violence was decreasing (Standal 2010). One man admitted that he no longer hit his wife. However, such evidence should be treated with caution since, although welcome, it is normative and does not necessarily signal a significant shift in gender relations. What people tell outsiders what they do and what they actually do in practice might not be the same thing.

Does access to modern energy services save time?

The answer to this question relates to context since other factors also influence time use. A review of the literature found that the amount of time spent on fuelwood collection varies, depending on site-specific environmental conditions, social arrangements and the distance to forest/ wasteland resources. In rural Zanzibar, despite the access to electricity the amount of time spent on fuelwood collection remained the same for some households. This fuelwood was used both for the household and a women's income generation project. Women need to walk further in search of wood due to trees becoming scarcer which is partly attributed to a tree-growing project that resulted in restricting access to land close to the village (Winther 2008). If a rural household has made the transition to kerosene or LPG, these fuels often need to be fetched from town which can take at least a day (Masera et al. 2000). Women in Sri Lanka cited the time saved on this trip is a benefit of electric cooking (Ramani and Heijndermans 2003).

In the kitchen, the time saved when switching to cleaner fuels appears to come from cleaning pots and pans, which are no longer caked in soot and ash, rather than from collecting fuelwood (Clancy et al. 2012).

What impact does modern energy access have on income poverty*?*

Poverty is considered a dynamic state, in the sense that households can move into or out of or deeper into poverty. Illness is considered the most overwhelming reason why households become poor primarily because a household member is too ill to work (CPRC 2008). Incapacity to work reduces the flow of money into the household but also if the ill person requires medicines probably household expenditure also increases due to the need to buy the medicines. Debt is another major cause of poverty. Poor people borrow money for a variety of reasons with the most cited are healthcare, investment in improving production (such as irrigation) and meeting social obligations particularly weddings and funerals (Krishna 2004).

Most of the attention has focused on electricity—in part because of the need to repay investment costs particularly in rural areas. There is general agreement that access to electricity is not a sufficient condition for income poverty reduction, that there are other complementary factors playing a role. A literature review by IDS and GIZ summarises these as: (1) access to finance for electric appliances (Khandker et al. 2012); (2) access to markets for the additional production (Peters et al. 2011); (3) skills for entrepreneurs to identify the new opportunities created by electricity and to prepare sound investment plans (Neelsen and Peters 2011); (4) access to other infrastructure or services (Kirubi et al. 2009); and (5) a high quality of the electricity supplied (Kooijman-van Dijk 2012; Kooijman-van Dijk and Clancy 2010) (IDS and GIZ 2019: 26). Reliability of supply is also found to be a key constraint to businesses with a negative impact on income due to delays in production, lost customers, lower sales and damaged expensive equipment (ENERGIA 2019). If the reaction by sufficiently large numbers of entrepreneurs is to buy individual generators or SHSs, it can significantly reduce the revenues of the power utility and its investment capacity to improve reliability (IDS and GIZ 2019). Nevertheless, SHSs while potentially more reliable than the grid do not offer the same level of power output which has implications for the equipment that can be used simultaneously.

Access to modern energy can impact on income poverty in two ways by reducing the household's overall costs and by increasing income. However, improved access alone is not enough to reduce income poverty. A range of complementary inputs, such as transport infrastructure, communication facilities, water supply and health and education services, are needed. Delivering a 'bundle' of other services together with energy provision can increase the benefits than if these services were delivered as separate projects (Peskin et al. 2000 cited in IDS 2003). For example, in Peru, analysis of the outcomes of adding a fourth service for rural households found that the effect was seven times greater than from the addition of the second service (Barnett 2000).

Indeed, access to electricity can contribute to increasing income poverty. In Zanzibar, the cost of a connection (as of the early 2000s) plus new appliances was reported as equivalent to five months of a fishermen income. Consequently, there was less money to buy necessities which was accompanied by a shift in responsibility to women for the households' daily living expenses (Winther 2008).

Access to more efficient technologies potentially reduces both energy poverty and can potentially help move households out of income poverty. A survey of poor households in Sri Lanka in the late 1990s found that savings of Rs100 on kerosene, estimated around 12% of monthly income, would have moved 22% above the poverty line (Gill et al. 1999). This of course assumes that households living poverty have sufficient financial resources to invest in energy-efficient technologies. Also, when a building needs attention for improved energy efficiency as relates to ambient temperatures, this can be problematic for households in rented accommodation since tenants are reliant on landlords/ladies making appropriate investments. This is a global problem. This responsibility may also encompass other domestic equipment such as for water heating and cooking.

In terms of contributing to improving household income, particularly for the anticipated opportunities modern energy is said to bring for women's economic opportunities, the evidence is mixed and depends on context. Econometric studies examining gender differences in employment opportunities linked to electricity found women's rate of employment increased by 9% in South Africa (Dinkelman 2011) and by 23% Nicaragua (Grogan and Sadanand 2013), while men's rate of employment in both contexts showed little change. A similar finding has been reported in the US (Lewis 2014). Care needs to be taken about generalising from these studies since there are other influencing factors that affect employment opportunities including the structure of the labour market which vary with context. An illustration of different employment outcomes due to electrification comes from India where a study found that only men benefitted from increased formal with a reduction in casual work by eight days, whereas women in the study's sample only benefitted by an extra four days per year of casual work (van de Walle et al. 2015).

3.10 Conclusions

Energy transitions can take time. A comprehensive survey in Hyderabad, India, on urban energy use found that it took at least 20 years for poor households to move from wood to kerosene. The shift was attributed to the middle classes moving to LPG and so reducing competition with the poor for kerosene (ESMAP 1999). In Addis Ababa, Ethiopia, again over a 20-year period (1980–2000) wood use reduced from being 70% of the total energy used to 13% (Shanko and Rouse 2005).

However, the literature does not include many longitudinal studies observing the social changes caused by electrification which makes it difficult to draw comprehensive conclusions. However, evidence does point to the factors which determine whether households adopt modern energy forms or technologies being more than simply availability and cost the assumptions that have since the 1980s shaped much policymaking. Several of the studies referred to in this chapter indicate that while there can be changes in gender roles as a consequence of access to modern energy services, this is not a given. Indeed, the available evidence related to changes in gender relations is not overwhelming. It is possible that these changes take time to emerge just like the changes in fuel used. Also, energy access alone is not sufficient to produce change, other contextual factors play a role.

There continue to be gaps in the data. One of the arguments for the promotion of alternatives to wood as a cooking fuel is the damage to a woman's body carrying the wood daily throughout her life. However, most of the evidence remains anecdotal (Matinga 2010). This means that serious health issues go unrecognised, non-prioritised and untreated.

Claims about changes to people's lives due to energy access should be treated with caution. Researchers report that statements made in response to questions about an issue, for example, the impact of electric light in households and what people claim to watch on TV do not always match with findings through participant observation (see, for example, Anneke undated; Matinga 2010). Indeed, in a review of the evidence, Ramani and Heijndermans considered that many of the reported claims about activities carried out when households acquire electric light are not possible with the level of lighting in respondents' homes (Ramani and Heijndermans 2003).

Context is also a key factor in outcomes which makes generalisation difficult. A perceived benefit of electric lighting in rural clinics is the contribution to improving safe deliveries. Women in Zanzibar saw another advantage: it saved money. Giving birth in your village, meant you did not need to spend on transportation to a more distant clinic, which could also entail having to spend several weeks with relatives awaiting your delivery (Winther 2008). However, in Tsilitwa, South Africa, women rejected giving birth in a well-equipped local clinic preferring to the hospital 35 km away which they evaluated as safer, as well as the social status acquired by having a private doctor (Matinga 2010).

Although understanding and addressing issues related to household energy started from different perspectives in the Global South and Global North, it is noticeable that there is now an overlap. While the focus in the Global South was on addressing access to modern energy sources, the assumption that this is not an issue in the Global North, has been found to be incorrect—a situation that the influence of the war in Ukraine on energy prices has brought to light. Households in the Global North are using wood fuels and peat and so are exposed to the levels of pollution, with the impacts on health, that are still unfortunately commonplace in the Global South. This opens the possibilities for South-North learning. Access to modern energy sources are promoted for a potential range of improvements to women's lives. However, the experiences of women in the Global North send a warning to women in the Global South that these improvements are not a given. For example, labour-saving equipment does not automatically translate into time-saving or a shift in gender roles. People own more clothes than previous generations as well as standards of cleanliness requiring more frequent washing. In Sweden, women spend more time on laundry than men (Carlsson-Kanyama and Linden 2007). In the UK, the 2017 British Social Attitudes survey reports that compared to 1987 there had only been a slight increase in men taking sole responsibility for the laundry (Scott and Clery 2013).

Development practice advocates that if women have access to labour-saving equipment, they will be able to use the time freed up for income generation. We explore in the next chapter, the role that energy access plays in enabling women to earn an income.

References

Abebaw, D. (2007). Household Determinants of Fuelwood Choice in Urban Ethiopia: A Case Study of Jimma. *The Journal of Developing Areas* 41: 117–126.

Agrawal, B. (1997). Gender, Environment and Poverty Interlinks: Regional Variations and Temporal Shifts in Rural India, 1971–91. *World Development* 25 (1): 23–52.

Ahmed, N. (2008). *Participatory Processes in Rural Energy Planning Under Energy Liberalisation Programmes in Sudan*. PhD Thesis, University of Twente, The Netherlands.

Andrade, T. (2004). *Enabling Urban Poor Livelihood Policy Making: Understanding the Role of Energy Services*. Brazil Country Report Prepared for

DFID KaR R8348. London: Department for International Development (DFiD).

Annecke, W.J. (1999). *Concept Paper for Energy and Women: Lessons Learned.* Cape Town: EDRC, University of Cape Town.

Annecke, W. (undated). *Whose Turn Is It to Cook Tonight? Changing Gender Relations in a South African Township.* Paper Prepared for the UK Department for International Development (DFID) KaR Research Project R8346, "Gender as a Key Variable in Energy Interventions". London: DFID.

Annecke, W. (2005). *An Assessment of PNES Customer Satisfaction and the Contribution of Electricity to the Quality of Life of Household in Khayelitsha, South Africa.* Report for EdF-ACCESS programme, Electricité de Francé, Paris.

Annecke, W. and Endelli, M. (2006). Gender and Prepayment Electricity in Merlo, Argentina. *Energia News* 9 (1): 18–20.

Approtech Asia. (2005). *Enabling Urban Poor Livelihood Policy Making: Understanding the Role of Energy Services.* Philippine Country Report Prepared for DFID KaR R8348. London: Department for International Development (DFiD).

Ato Armah, F., Ekumah, B., Yawson, D.O., Odoi, J.O., Afitiri, A.-R. and Nyieku, F.E. (2019). Predictive Probabilities of Access to Clean Cooking: Evidence From the Demographic and Health Surveys in 31 Countries in Sub-Saharan Africa. *Environmental Justice* 12 (3): 118–131.

Bannister, A. (2002). The Sustainable Urban Livelihoods Framework—A Tool for Looking at the Links Between Energy and Poverty. *Boiling Point* 48: 710.

Barbosa Neves, B., Amaro, F. and Fonseca, J.R.S. (2013). Coming of (Old) Age in the Digital Age: ICT Usage and Non-Usage among Older Adults. *Sociological Research Online* 18 (2): 22–35.

Barkat, A., Khan, S.H., Rahman, M., Zaman, S., Poddar, A., Halim, S., Ratna, N.H., Majid, M., Maksud, A.K.M, Karim, A. and Islam, S. (2002). *Economic and Social Impact Evaluation Study of the Rural Electrification Program in Bangladesh.* Dhaka: Human Development Research Center (HDRC), NRECA International Ltd, Rural Electrification Board of Bangladesh and USAID for the Rural Power for Poverty Reduction Program.

Barnes, D.F. and Sen, M. (2004). *The Impact of Energy on Women's Lives in Rural India.* Washington, DC: ESMAP, The World Bank.

Barnett, A. (2000). *Energy and the Fight Against Poverty.* Paper Presented as Part of Series of Economic Research Seminars, 29th June, Institute of Social Studies, The Hague.

Bates, E. (2009). Making LPG Stoves Accessible for Low-Income Communities in Kassala, Sudan. In Rai, K. and McDonald, J.U. (eds.), *Cookstoves and Markets: Experiences and Success and Opportunities*. London: GVEP International.

Behsudi, A. (2021). *Radical Inclusion*. Washington, DC: Finance and Development, International Monetary Fund. https://www.imf.org/external/pubs/ft/fandd/2021/03/david-sengeh-of-sierra-leone-on-digital-educationand-economy-trenches.htm (accessed 23 June 2023).

Belmin, C., Hoffmann, R., Pichler, P-P. and Weisz, H. (2022). Fertility Transition Powered by Women's Access to Electricity and Modern Cooking Fuels. *Nature Sustainability* 5: 245–253.

Birgi, O.G., Fuhrmann, A., Habersbrunner, K. and Stock, A. (2021). *Gender and Energy Poverty—Facts and Arguments*. Report Prepared for EmpowerMed. https://www.empowermed.eu/wp-content/uploads/2021/05/2104.Empowermed-Energy_Poverty_and_gender.pdf (accessed 10 October 2022).

Blackden, C.M. and Wodon, Q. (eds.). (2006). *Gender, Time Use, and Poverty in Sub-Saharan Africa*. World Bank Working Paper No. 73. Washington, DC: The World Bank.

Boardman, B. (1991). *Fuel Poverty: From Cold Homes to Affordable Warmth*. London: Bellhaven.

Bouzarovski, S. and Petrova, S. (2015). A Global Perspective on Domestic Energy Deprivation: Overcoming the Energy Poverty–Fuel Poverty Binary. *ERSS* 10: 31–40.

Bouzarovski, S. and Tirado Herrero, S. (2014). The Energy Divide: Integrating Energy Transitions, Regional Inequalities and Poverty Trends in the European Union. *European Urban and Regional Studies* 24 (1): 69–86.

Brida, M. (undated). *Why Do Women Feel the Cold More Than Men?* https://guysandstthomasspecialistcare.co.uk/news/why-do-women-feel-the-cold-more-than-men/ (accessed 17 May 2024).

Bruce, N., Rehfuess, E., Mehta, S., Hutton, G. and Smith, K. (2006). Indoor Air Pollution. In Jamison, D.T., Breman, J.G., Measham, A.R., Alleyne, G., Claeson, M., Evans, D.B., Jha, P., Mills, A. and Musgrove, P. (eds.), *Disease Control Priorities in Developing Countries*. Washington, DC: World Bank.

Butchart, A. (2000). *A Profile of Fatal Injuries in South Africa 1999: First Annual Report of the National Injury Mortality Surveillance System*. The Violence and Injury Surveillance Consortium, with Participating Forensic Pathologists and the State Forensic Chemistry Labor. Cape Town: MRC/UNISA.

CARE. (1999). *Household Livelihood Security in Urban Settlements*. CARE International UK. Urban Briefing Notes. December (1).

Carlsson-Kanyama, A. and Linden, A-L. (2007). Energy Efficiency in Residences—Challenges for Women and Men in the North. *Energy Policy* 35: 2163–2172.

Castán Broto, V., Arthur, M.S.R. and Guibrunet, L. (2020). Energy Profiles Among Urban Elite Households in Mozambique: Explaining the Persistence of Charcoal in Urban Areas. *Energy Research & Social Science* 65: 101478.

Castaño-Rosa, R. and Okushima, S. (2021). Prevalence of Energy Poverty in Japan: A Comprehensive Analysis of Energy Poverty Vulnerabilities. *Renewable and Sustainable Energy Reviews* 145: 111006.

Cecelski, E. (2000). *The Role of Women in Sustainable Energy Development*. NREL/SR-550-26889. Golden, Colorado, USA: National Renewable Energy Laboratory (NREL).

Cecelski, E. (2004). *Re-thinking Gender and Energy: Old and New Directions*. ENERGIA/EASE Discussion Paper. www.energia.org

Cecelski, E. (2005). *Energy, Development and Gender: Global Correlations and Causality*. Report Prepared as Part of the United Kingdom Department for International Development (DfID) KaR Research Project R8346 on "Gender as a Key Variable in Energy Interventions." www.energia.org (accessed 8 March 2023).

Cecelski, E., Ounali, A., Aissa, M. and Dunkerley, J. (2005). *Rural Electrification in Tunisia: National Commitment, Efficient Implementation, and Sound Finances*. Washington, DC: ESMAP, The World Bank.

Chaieb, S. and Ounalli, A. (2001). Rural Electrification Benefits Women's Health, Income and Status in Tunisia. *ENERGIA News* 4 (4): 18–20.

Chakrabarti, S. and Chakrabarti, S. (2002). Rural Electrification Programme With Solar Energy in Remote Region—A Case Study in an Island. *Energy Policy* 30 (1): 33–42.

Chard, R. and Walker, G. (2016). Living With Fuel Poverty in Older Age: Coping Strategies and Their Problematic Implications. *Energy Research & Social Science* 18: 62–70.

Charmes, J. (2006). *A Review of Empirical Evidence on Time Use in Africa from UN-Sponsored Surveys*. In Blackden, C.M. and Wodon, Q. (eds.), Gender, Time Use, and Poverty in Sub-Saharan Africa. World Bank Working Paper No. 73. Washington, DC: The World Bank.

Citizens Advice. (2023). *Millions Left in the Cold and Dark as Someone on a Prepayment Meter Cut Off Every 10 Seconds, Reveals Citizens Advice*. https://www.citizensadvice.org.uk/about-us/about-us1/media/press-releases/millions-left-in-the-cold-and-dark-as-someone-on-a-prepayment-meter-cut-off-every-10-seconds-reveals-citizensadvice (accessed 26 June 2023).

Clancy, J.S. and Roehr, U. (2003). Gender and Energy: Is There a Northern Perspective? *Journal of Energy for Sustainable Development* VII (3): 44–50.

Clancy, J., Daskalova, V., Feenstra, M., Franceschelli, N. and Sanz, M. (2017). *Gender Perspective on Access to Energy in the EU*. Study for the FEMM Committee of the EU Parliament, PE 596.816, Brussels, 2017.

Clancy, J.S., Maduka, O. and Lumampao, F. (2006). Sustainable Energy Systems and Urban Poor Livelihoods. In Droege, P. (ed.), *Urban Energy Transition: From Fossil Fuels to Renewable Power*. Elsevier.

Clancy, J., Winther, T., Matinga, M. and Oparaocha, S. (2012) *Gender Equity in Access to and Benefits From Modern Energy and Improved Energy Technologies*. World Development Report 2012 Background Paper. Washington, DC: World Bank.

Cooke, P., Köhlin, G. and Hyde, W.F. (2008). Fuelwood, Forests and Community Management—Evidence From Household Studies. *Environment and Development Economics* 13: 103–135.

CPRC. (2008). *The Chronic Poverty Report 2008–09: Escaping Poverty Traps*. Manchester: The Chronic Poverty Research Centre (CPRC).

Davis, M. (1998). Rural Household Energy Consumption: The Effects of Access to Electricity Evidence From South Africa. *Energy Policy* 26(3): 207-217.

Dhanapala, K. (1995). *Report on the Gender Related Impact of Micro Hydro Technology at the Village Level*. Study Report Number 2. Colombo, Sri Lanka: Intermediate Technology.

Dinkelman, T. (2011). The Effects of Rural Electrification on Employment: New Evidence From South Africa. *American Economic Review* 101 (7): 3078–3108.

Dutta, S. (2005). *Energy as a Key Variable in Eradicating Extreme Poverty and Hunger: A Gender and Energy Perspective on Empirical Evidence on MDG #1*. Discussion Paper Prepared for UK Department for International Development (DFID) KaR Research Project R8346 "Gender as a Key Variable in Energy Interventions". The Hague: ENERGIA/HIVOS.

Eakins, J., Power, B., Dunphy, N. and Sirr, G. (2022). *Residential Solid Fuel Use in Ireland and the Transition Away from Solid Fuels*. Report Prepared for the Environmental Protection Agency. ISBN: 978-1-80009-035-4. Wexford, Ireland: Environmental Protection Agency.

Echarri, J.J. and Forriol, F. (2002). Effect of Axial Load on the Cervical Spine: A Study of Congolese Woodbearers. *International Orthopaedics* 26: 141–144.

EIGE. (2020). *Area A—Women and Poverty: Women at Greater Risk*. Report as Part of the Fifth Review of the Implementation of the Beijing Platform for Action (Beijing + 25) in the EU Member States. Vilnius, Lithuania: The European Institute for Gender Equality (EIGE).

EIGE. (2022). *Gender Inequalities in Informal Long-Term Care*. Statistical Brief, 2022. Vilnius, Lithuania: The European Institute for Gender Equality (EIGE).

Elnakat, A. and Gomez, J.D. (2015). Energy Engenderment: An Industrialized Perspective Assessing the Importance of Engaging Women in Residential Energy Consumption Management. *Energy Policy* 82: 166–177.

ENERGIA. (2006). *From the Millennium Development Goals: Towards a Gender-Sensitive Energy Policy Research and Practice: Empirical Evidence and Case Studies*. Synthesis Report for UK Department for International Development (DFID) KaR Research Project R8346 "Gender as a Key Variable in Energy Interventions". The Hague: ENERGIA/HIVOS.

ENERGIA. (2019). *Gender in the Transition to Sustainable Energy for All: From Evidence to Inclusive Policies*. Synthesis Report of the Evidence Generated by the ENERGIA Gender and Energy Research Programme. The Hague: ENERGIA/HIVOS. https://www.energia.org/new-gender-in-the-transition-to-energy-for-all-from-evidence-to-inclusive-policies/ (accessed 9 March 2022).

EPSECC (Environmental Policy, Social Exclusion and Climate Change Research Co-operation). (1997). *Environmental Policy, Social Exclusion and Climate Change Research Co-operation*. Work Package 2: Data Analysis.

ESG and CSD-NGO WC (ENERGIA Support Group and CSD-NGO Women's Caucus). (2001). *Gender Perspectives on Energy for CSD-9*. Position Paper Prepared for the 9th Session of the Commission on Sustainable Development. http://www.energia.org

ESMAP. (1999). *Household Energy Strategies for Urban India; the Case of Hyderabad*. Report 214/99. Washington, DC: The World Bank.

European Commission. (2020). *Striving for a Union of Equality: The Gender Equality Strategy 2020–2025*. Luxembourg: Publications Office of the European Union.

Eurostat. (2021). *Digital Economy and Society Statistics—Households and Individuals*. https://ec.europa.eu/eurostat/statisticsexplained/index.php?title=Digital_economy_and_society_statistics_-_households_and_individuals (accessed 23 August 2022).

Fernández-Baldor, A., Lillo, P. and Boni, A. (2015). Gender, Energy, and Inequalities: A Capabilities Approach Analysis of Renewable Electrification Projects in Peru. In Hostettler, S., Gadgil, A. and Hazboun, E. (eds.), *Sustainable Access to Energy in the Global South. Essential Technologies and Implementation Approaches* (pp. 193–204). Springer International.

Fouillet, A., Rey, G., Laurent, F., Pavillon, G., Bellec, S., Guihenneuc-Jouyaux, C., Clavel, J., Jougla, E. and Hémon. D. (2006). Excess Mortality Related to the August 2003 Heatwave in France. *International Archives of Occupational and Environmental Health* 80 (1): 16–24.

GACC. (2016). *Progress report 2016*. Washington, DC: Global Alliance for Clean Cookstoves (GACC).

Gill, A., Mosley, P. and Fulford, D. (1999). *Community Micro-hydro in LDCs: Adoption, Management and Poverty Impact*. ESD and University of Reading. Project 7110. London: Department for International Development.

Green, M.G. (2001). *Solar Cookers as a Mechanism for Women's Empowerment.* Paper Presented at ISES World Solar Congress, Adelaide, Australia, December 2001.

Grimm, M., Sparrow, R. and Tasciotti, L. (2015). Does Electrification Spur the Fertility Transition? Evidence From Indonesia. *Demography* 52 (5): 1773–1796.

Grogan, L. and Sadanand, A. (2013). Rural Electrification and Employment in Poor Countries: Evidence From Nicaragua. *World Development* 43 (C): 252–265.

GSMA. (2010). *Women & Mobile: A Global Opportunity: A Study on the Mobile Phone Gender Gap in Low and Middle-Income Countries.* GSMA Development Fund and Cherie Blair Foundation for Women. https://cherieblairfoundation.org/what-we-do/research/women-and-mobile-a-global-opportunity/ (accessed 16 August 2022).

GSMA. (2020). *The Mobile Gender Gap Report.* https://www.gsma.com/r/gender-gap/ (accessed 12 August 2022).

GTZ. (1999). *Moving Ahead With Solar Cookers: Acceptance and Introduction to the Market.* Eschborn, Germany: Deutsche GesellschaGft für Technische Zusammenarbeit (GTZ) GmbH.

Haile, F. (1989). Women Firewood Carriers and the Supply of Household Energy in Addis Ababa. *Canadian Journal of African Studies* 23: 442–451.

Haile, F. (1991). *Women Fuelwood Carriers in Addis Ababa and the Peri-Urban Forest.* Geneva: ILO.

Hawthorne, W. (1996). *Women and Large-Scale Electricity Development.* Working Paper #257. East Lansing, MI, US: Michigan State University.

HDRC. (2002). *Economic and Social Impact Evaluation Study of the Rural Electrification Program.* Human Development Research Centre, for NRECA International Ltd's Rural Power for Poverty Reduction Program (USAID CA388-A-00-97-0040-00), Dhaka, Bangladesh. October 2002.

IDS. (2003). *Energy, Poverty Gender: A Review of the Evidence and Case Studies in Rural China.* Washington DC: The World Bank.

IDS and GIZ. (2019). *Unlocking the Benefits of Productive Uses of Energy for Women in Ghana, Tanzania and Myanmar.* Research Report RA6, ENERGIA Gender and Energy Research Programme. The Hague: ENERGIA/HIVOS.

IEG. (2008). *The Welfare Impact of Rural Electrification: A Reassessment of the Costs and Benefits.* Washington, DC: The World Bank Independent Evaluation Group (IEG).

IISD (Global Subsidies Initiative), BIDS, IRADe and Spaces for Change. (2019). *Gender and Fossil Fuel Subsidy Reform: Current Status of Research.* Research report RA4. ENERGIA Gender and Energy Research Programme. The Hague: ENERGIA/HIVOS.

Kasirye, B.G., Matinga, M. and Clancy, J. (2009). Fuel Security and Supply Dynamics in Internally Displaced Persons' Camps of Northern Uganda. *Journal of Humanitarian Assistance*. http://sites.tufts.edu/jha/archives/462

Kelkar, G. and Nathan, D. (2005). *Gender Relations and the Energy Transition in Rural Asia*. Paper Prepared for Department for International Development (DFID) KaR Research Project R8346, "Gender as a Key Variable in Energy Interventions".

Kelkar, G. and Nathan, D. (2007). *Testing of Gender Tools for Energy Projects in India*. Report for ENERGIA. Leusden, The Netherlands: ENERGIA.

Khandker, S.R., Samad, H.A., Ali, R. and Barnes, D.F. (2012). *Who Benefits Most From Rural Electrification? Evidence in India*. World Bank Policy Research Working Paper 6095. Washington, DC: The World Bank.

Kirubi, C., Jacobson, A., Kammen, D.M. and Mills, A. (2009). Community-Based Electric Micro-Grids Can Contribute to Rural Development: Evidence from Kenya. *World Development* 37 (7): 1208–1221.

Kisyov, P. (2014). Report on National Situation in the Field of Fuel Poverty: Bulgaria. REACH Research Programme Under the Intelligent Energy Europe Programme of the European Union. http://reach-energy.eu/wordpress/wpcontent/uploads/2014/12/D2.2-EAP_EN.pdf (accessed 26 June 2023).

Köhlin, G., Sills, E.O., Pattanayak, S.K. and Wilfong, C. (2011). *Energy, Gender and Development: What Are the Linkages? Where Is the Evidence?* Policy Research Working Paper 5800—Background Paper to the 2012 World Development Report. Washington, DC: Social Development Unit, World Bank.

Kooijman-van Dijk, A.L. (2012). The Role of Energy in Creating Opportunities for Income Generation in the Indian Himalayas. *Energy Policy* 41: 529–36.

Kooijman-van Dijk, A.L. and Clancy, J. (2010). Impacts of Electricity Access to Rural Enterprises in Bolivia, Tanzania and Vietnam. *Energy for Sustainable Development* 14 (1): 14–21.

Krishna, A. (2004). Escaping Poverty and Becoming Poor: Who Gains, Who Loses and Why? *World Development* 32: 121–136.

Lahn, G. and Grafham, O. (2015). *Heat, Light and Power for Refugees: Saving Lives, Reducing Costs*. Chatham House Report for the Moving Energy Initiative. Chatham House, London: The Royal Institute of International Affairs.

Lewis, J. (2014). *The Impact of Technological Change Within the Home*. PhD Thesis, University of Toronto, Canada.

Lim, S.S. et al. (2010). A Comparative Risk Assessment of Burden of Disease and Injury Attributable to 67 Risk Factors and Risk Factor Clusters in 21 Regions 1990–2010: A Systematic Analysis for the Global Burden of Disease Study 2010. *The Lancet* 380, iss. 9859 (15 December 2010): 2224–2260.

Lovins, A.B. (1976). Energy Strategy: The Road Not Taken. *Foreign Affairs* 55 (1): 65.

Lumampao, F.G., Mataga, R.S. and Parado, B.C. (2006). Energising Poor Women Entrepreneurs. *ENERGIA News* 9 (2): 18–20.

Mahat, I. (2004). *Integrating Gender into Planning Management and Implementation of Rural Energy Technologies: The Perspectives of Women in Nepal*. Ph.D. thesis submitted to Institute of Development Studies, School of People, Environment and Planning, Massey University, New Zealand.

Makan, A. (1995), Power for Women and Men: Towards a Gendered Approach to Domestic Policy and Planning in South Africa. *Third World Planning Review* 17 (2): 183–198.

Maleko, G.C. (2006). *Impact of Electricity Services on Microenterprise in Rural Areas in Tanzania*. Master's Thesis, University of Twente, Enschede, The Netherlands.

Malmberg Calvo, C. (1994). *Case Study on the Role of Women in Rural Transport: Access of Women to Domestic Facilities*. Sub-Saharan Africa Transport Policy Program. Washington, DC: World Bank.

Mandhlazi, W. (1999). *A Reflection on the Impact of Renewable Energy Projects on Gender*. SAREIN News Flash, June.

Masera, O.R., Saatkamp, B.D. and Kammen, D.M. (2000). From Linear Fuel Switching to Multiple Cooking Strategies: A Critique and Alternative to the Energy Ladder Model. *World Development* 28 (12): 2083–2103.

Massé, R. (2003). *Impacts of Rural Electrification on Poverty and Gender in Sri Lanka*. Labastide-Murat, France: MARGE.

Matinga, M.N. (2008). The Making of Hardiness in Women's Experience of Health Impacts of Wood Collection and Use in Cuntwini, Rural South Africa. *Medische Anthropologie* 20: 279–295.

Matinga, M.N. (2010). *We Grow Up With It: An Ethnographic Study of the Experiences, Perceptions and Responses to the Health Impacts of Energy Acquisition and Use in Rural South Africa*. PhD thesis, University of Twente, Enschede, The Netherlands.

Matinga, M.N. (2012). *Gender Equity in Access to and Benefits From Modern Energy and Improved Energy Technologies*. Background Paper Prepared for the 2012 World Development Report. ETC/ENERGIA in Association Nord/Sør-konsulentene.

Matinga, M. and Clancy, J. (2020). Gender, Firewood and Health: The Potential of Ethnography to Inform Policy and Practice. In Clancy, J., Őzeröl, G., Mohlakoana, N., Feenstra, M. and Sol Cueva, L. (eds.), *Engendering the Energy Transition*. Palgrave Macmillan.

Matly, M. (2003). *Rural Electrification in Indonesia and Sri Lanka: From Social Analysis to Reform of the Power Sector*. ASTAE EnPoGen Project. Washington, DC: The World Bank.

Mehretu, A. and Mutambira, C. (1992). Gender Differences in Time and Energy Costs of Distance for Regular Domestic Chores in Rural Zimbabwe. *World Development* 20 (11): 1675–1683.

Meikle, S. (2004). *A Study of the Impact of Energy Use on Poor Urban Women and Girls' Livelihoods in Arusha, Tanzania*. London: University College London.

Moniruzzaman, M. and Day, R. (2020). Gendered Energy Poverty and Energy Justice in Rural Bangladesh. *Energy Policy* 144: 111554.

MSF. (2005). *The Crushing Burden of Rape: Sexual Violence in Darfur*. Amsterdam: Médecins sans Frontières.

MSSRF and CRT Nepal. (2016). *The Gender Factor in Political Economy of Energy Sector Dynamics*. Scoping Study Report RA3 Prepared for the ENERGIA Gender and Energy Research Programme 'Building the Evidence Base for Improving Energy Interventions' Effectiveness by Taking'. www.energia.org. The Hague: ENERGIA/HIVOS.

MSSRF and CRT Nepal. (2019). *The Gender Factor in Political Economy of Energy Sector Dynamics*. Research Report RA3 Prepared for the ENERGIA Gender and Energy Research Programme 'Building the Evidence Base for Improving Energy Interventions' Effectiveness by Taking'. www.energia.org. The Hague: ENERGIA/HIVOS.

Muller, C. and Yan, H. (2018). Household Fuel Use in Developing Countries: Review of Theory and Evidence. *Energy Economics* 70: 429–439.

Mulokozi, G., Mselle, L., Mgoba, C., Mugyabuso, J.K.L., and Ndossi, G.D. (2000). *Improved Solar Drying of Vitamin A-Rich Foods by Women's Groups in the Singida District of Tanzania*. Nairobi, Kenya: International Center for Research on Women.

Nathan, D. and Kelkar, G. (1997). Wood Energy: The Role of Women's Unvalued Labor. *Gender, Technology and Development* 1 (2): 205–224.

Neelsen, S. and Peters, J. (2011). Electricity Usage in Micro-Enterprises—Evidence From Lake Victoria, Uganda. *Energy for Sustainable Development* 15 (1): 21–31.

Novoa, A. and Bottazzi, L. (2018). *Radiografies de la situació del dret a l'habitatge, la pobresa energètica i el seu impacte en la salut a Barcelona*. Aliança contra la Pobresa Energètica, Observatori DESC, Plataforma d'Afectades per la Hipoteca, Enginyeria sense Fronteres.

Nyoni, S. (1993). *Women and Energy: Lessons From the Zimbabwe Experience*. Working paper 22. Harare: Zimbabwe Environment Research Organisation (ZERO).

ONPE. (2016). *Les chiffres clés de la précarité énergétique*. Edition 2, Novembre 2016: http://www.onpe.org/sites/default/files/pdf/tableau_de_bord/chiffrescles-precarite-energetique-novembre2016.pdf (accessed 1 May 2019).

Ouedraogo, B. (2006). Household Energy Preferences for Cooking in Urban Ouagadougou, Burkina Faso. *Energy Policy* 34: 3787–3795.

Palit, D., Bhattacharyya, S.C. and Chaurey, A. (2014). Indian Approaches to Energy Access. In Halff, A., Sovocol, B. and Rozhon, J. (eds.), *Energy Poverty: Global Challenges and Local Solutions* (pp. 237–256). Oxford: Oxford University Press.

Peskin, H., Barnes, D., Domdom, A. and de Gui-Abiad, V. (2000). *New Approaches to Evaluate Rural Electricity Projects*. Presentation September 27. Washington, DC: World Bank.

Peters, J., Vance, C. and Harsdorff, M. (2011). Grid Extension in Rural Benin: Micro-Manufacturers and the Electrification Trap. *World Development* 39 (5): 773–783.

Petrova, S. (2017). Illuminating Austerity: Lighting Poverty as an Agent and Signifier of the Greek Crisis. *European Urban and Regional Studies* 1–13.

Porcaro, J. and Takada, M. (eds.). (2005). *Achieving the Millennium Development Goals: The Role of Energy Services*. New York: United Nations Development Program.

Porter, G., Hampshire, K., Dunn, C., Hall, R., Levesley, M., Burton, K., Robson, S., Abane, A., Blell, M. and Panther, J. (2013). Health Impacts of Pedestrian Head-Loading: A Review of the Evidence with Particular Reference to Women and Children in sub-Saharan Africa. *Social Science and Medicine* 88: 90–97.

Potgieter, C.-A., Pillay, R. and Rama, S. (2006). *Women, Development and Transport in Rural Eastern Cape, South Africa*. Cape Town: HSRC Press.

Preisendoerfer, P. (1999). *Umwelteinstellungen und Umweltverhalten in Deutschland: Empirische Befunde und Analysen auf der Grundlage der Bevoelkerungsumfragen "Umweltbewusstsein in Deutschland 1991–1998"*. Herausgegeben vom Umweltbundesamt (Environmental Attitudes and Environmental Behaviour in Germany), Leske und Budrich, Opladen.

Putti, V.R, Tsan, M., Mehta, S. and Srilata, K. (2015). *The State of the Global Clean and Improved Cooking Sector*. ESMAP Technical Paper No. 007/15. Washington, DC: The World Bank. https://openknowledge.worldbank.org/handle/10986/21878

Puzzolo, E., Stanistreet, D., Pope, D., Bruce, N. and Rehfuess, E. (2013). *Factors Influencing the Largescale Uptake by Households of Cleaner and More Efficient Household Energy Technologies*. London: EPPI-Centre, Social Science Research Unit, Institute of Education, University of London.

Rahuta, D.B., Das, S., De Groote, H. and Behera, B. (2014). Determinants of Household Energy Use in Bhutan. *Energy* 69: 661–672.

Ramani, K.V. and Heijndermans, E. (2003). *Energy, Poverty and Gender: A Synthesis*. Washington, DC: The World Bank.

Rodgers, P.M. (1994). *Solar Box Cookers in Zimbabwe. The Introduction of Radical Innovation in Cooking*. Unpublished MSc Thesis, University of California, Davis. http://solarcooking.org/zimb-the.htm (accessed 8 March 2022).

Rollin, H., A. Mathee, A., Bruce, N.G., Levin, J. and von Schirnding, Y.E.R. (2004). Comparison of Indoor Air Quality in Electrified and Un-Electrified Dwellings in Rural South African Villages. *Indoor Air* 14 (3): 208–216.

Rossi, A. and Lambrou, Y. (2008). *Gender and Equity Issues in Liquid Biofuels Production: Minimizing the Risks to Maximize the Opportunities*. Rome: Food and Agriculture Organisation (FAO).

Saptyani, G. (2010). *Kerosene to LPG Conversion Program in Indonesia: Evaluation of the Effects on Women's Livelihoods*. Master's Thesis, University of Twente, Enschede, The Netherlands.

Scott, J. and Clery, E. (2013). Gender Roles. In Park, A., Bryson, C., Clery, E., Curtice, J. and Phillips, M. (eds.), *British Social Attitudes: The 30th Report*. London: NatCen Social Research. www.bsa-30.natcen.ac.uk (accessed 27 July 2017).

SEforALL. (2018). *Levers of Change: How Global Trends Impact Gender Equality and Social Inclusion in Access to Sustainable Energy*. Vienna, Austria: Sustainable Energy for All (SEforAll).

Shanko, M. and Rouse, J. (2005). The Human and Livelihoods Cost of Fuel-Switching in Addis Ababa. *Boiling Point* 51: 31–33.

Snell, C., Bevan, M. and Thomson, H. (2015). Justice, Fuel Poverty and Disabled People in England. *Energy Research & Social Science* 10: 123–132.

Soria, L., Farley, K. and Glinski, A. (2016). *With Solar Sister, Forward We Go: A Qualitative Assessment Exploring How Solar Sister Brings Light, Hope, and Opportunity to Women in Africa*. Washington, DC: International Center for Research on Women. https://www.icrw.org/wpcontent/uploads/2016/10/Solar-Sister-Qualitative-Assessment.pdf

Sovacool, B.K., Cooper, C., Bazilian, M., Johnson, K., Zoppo, D., Clarke, S., Eidsness, J., Crafton, M., Velumail, T. and Raza, H.A. (2012). What Moves and Works: Broadening the Consideration of Energy Poverty. *Energy Policy* 42: 715–719.

Standal, K. (2010). *Giving Light and Hope in Rural Afghanistan: Enlightening Women's Lives With Solar Energy*. Lampert Academic Publishing.

Standal, K. (2018). *Challenges of Gender, Power and Change in Solar Energy Interventions in Rural India*. PhD Thesis, University of Oslo, Norway.

Suri, T. and Jack, W. (2016). The Long-Run Poverty and Gender Impacts of Mobile Money. *Science* 354: 1288–1292.

Thomson, H., Snell, C. and Bouzarovski, S. (2017). Health, Well-Being and Energy Poverty in Europe: A Comparative Study of 32 European Countries. *International Journal of Environmental Research and Public Health* 14: 584–604.

Tucker, M. (1999). Can Solar Cooking Save Forests? *Ecological Economics* 31: 77–89.

UNIDO-UN Women (United Nations Industrial Development Organization-UN Women). (2013). *Sustainable Energy for All: The Gender Dimensions*. New York: UN Women.

United Nations Millennium Development Project. (2005). *Investing in Development: A Practical Plan to Achieve the Millennium Development Goals*. Report to the UN Secretary-General. New York: United Nations Development Programme.

University of Oslo, TERI, Seacrester Consulting and Dunamai Energy. (2019). *Women's Empowerment and Electricity Access: How Do Grid and Off-Grid Systems Enhance or Restrict Gender Equality?* Research report RA1, ENERGIA Gender and Energy Research Programme. The Hague: ENERGIA/HIVOS.

van de Walle, D., Ravallion, M., Mendiratta, V. and Koolwal, G. (2015). Long-Term Gains From Electrification in Rural India. *The World Bank Economic Review*, October 2015, 1–36. https://doi.org/10.1093/wber/lhv057

Vogt, K.A. (2007). *Forests and Society: Sustainability and Life Cycles of Forests in Human Landscapes*. CABI.

Wamukonya, L. and Davis, M. (1999). *Socio-Economic Impacts of Rural Electrification in Namibia*. Report 1: Comparisons Between Grid, Solar and Unelectrified Households. EDRC, University of Cape Town, South Africa.

WHO. (2014). *Burden of Disease From Household Air Pollution for 2012*. Geneva: World Health Organisation (WHO).

Wickramasinghe, A. (2003). Gender and Health Issues in the Biomass Energy Cycle: Impediments to Sustainable Development. *Energy for Sustainable Development* VII: 51–61.

Wilson, M. (1961). *A Reaction to Conquest: Effects of Contact With Europeans on the Pondo of South Africa*. Oxford: Oxford University Press.

Winther, T. (2008). *The Impact of Electricity. Development, Desires and Dilemmas*. Oxford: Berghahn Books.

Winther, T. (2012). *Electricity's Effect on Gender Equality in Rural Zanzibar, Tanzania*. Case Study for Gender and Energy World Development Report Background Paper. World Development Report 2012 Background Paper. Washington, DC: World Bank.

Winther, T. (2014). *The Introduction of Electricity in the Sundarban Islands: Conserving or Transforming Gender Relations?* In Nielsen, K.B. and Waldrop, A. (eds.), Women, Gender and Everyday Social Transformation in India. London: Anthem Press.

Winther, T., Palit, D., Govindan, M., Matinga, M., Standal, K., Ulsrud, K., Saini, A. and Gichungi, H. (2016). *Exploring Factors That Enhance and Restrict Women's Empowerment Through Electrification (EFEWEE)*. Scoping Report Prepared for the ENERGIA Gender and Energy Research Programme

'Building the Evidence Base for Improving Energy Interventions' Effectiveness by Taking a Gender Approach'. www.energia.org. The Hague: ENERGIA.

World Bank. (2005). *World Development Indicators 2005*. Washington, DC: World Bank.

World Bank. (2012). *State of the Clean Cooking Energy Sector in Sub-Saharan Africa*. Washington, DC: World Bank.

World Bank. (2023). *Tracking SDG7: The Energy Progress Report 2023*. Washington, DC: World Bank.

World Bank and IEA. (2015). *Global Tracking Framework 2015*. Washington, DC: World Bank.

CHAPTER 4

Women, Energy and Work

In this chapter we look at two different aspects in the way that the energy system influences women's capacity to earn an income: by analysing energy as an input into women's enterprises and the energy sector as a source of employment. There is quite an extensive academic and grey literature which focuses on the outcomes of women's employment. An instrumental argument for promoting women's income generation is that women are considered to invest more of their earnings in their family than men resulting in reducing household poverty and improved health and education opportunities for their children (Kabeer 2012). Evidence from Rwanda supports this notion indicating that women use their income from businesses primarily for food for the household whereas men buy alcohol and livestock with some put aside as savings (IDS and GIZ 2019). While evidence from the Dominican Republic found that men entrepreneurs were earning about 20% more than women entrepreneurs from their enterprises (which in part reflects the nature of their enterprises), as well as having other more lucrative sources of income, while they contribute about the same amount to their households (Espinal and Grasmuck 1997). When analysing the impacts of energy access for small enterprises, most of the available evidence tends to relate to women in the Global South with little attention to men's energy use. Analysis is not generally from a gender perspective but with women as the focus—hence the title of this chapter. The rural household has been the

J. Clancy, *Engendering the Energy System*,
https://doi.org/10.1007/978-3-031-78965-6_4

primary unit of analysis (IDS and GIZ 2019), although there has been interest in the urban informal sector (Clancy et al. 2006; University of Twente et al. 2015). There is very little evidence from a gender perspective about energy as a factor of input into enterprises related to the Global North where the focus has been on gender disparity in employment in the commercial energy sector.

If an intended outcome of modern energy access is to support existing and create new productive activities which generate income, then the objective has a gender dimension since women are less likely than men to be employed, women's wages, even in like for like jobs, are more likely to be lower than men's and women are found in a narrow range of low-productivity occupations and low-remuneration rates (Addati et al. 2016; Campos et al. 2015; Morton et al. 2014; de Haan 2016; Cirera and Qasim 2014). In terms of business ownership, globally, men's ownership or management of a business is estimated at twice as likely compared to women (GEM 2018), while one-third of formal businesses are completely or partly owned by women (Kelley et al. 2017) with regional variations (for example, Uruguay 2%, Europe 5.7%, Angola 9.9%, Iran 6.6%) linked to opportunities in other forms of employment or social norms related to women working (GEM 2018). The available evidence indicates that at start-up women are much more likely than men to be self-employed (Johns Hopkins University et al. 2019). It appears, at least in the context of Austria, that women business owners are more likely than their male counterparts to hire women (Weber and Zulehner 2010). Evidence shows that women business owners tend to perform as well or better than their male counterparts operating in similar businesses (Du Reitz and Henrekson 2000; Robb and Watson 2011; Zolin et al. 2013; Shankar et al. 2015; Brush and Elam 2023).

The chapter starts in Section 4.1 with an overview of what it means to be an entrepreneur pointing out some of the gender influences on the nature of the options for enterprises particularly for women. Section 4.2 describes the limited evidence about the role of energy in women's enterprises providing an income while Section 4.3 focuses on women working in the energy sector from two perspectives: running energy enterprises and as employees. Section 4.4 looks at how the energy sector could be more gender-equitable and socially inclusive. Section 4.5 reviews the evidence about the outcomes for women working in the energy sector.

The chapter closes with a section that draws some conclusions as to whether expectations about women working in the energy sector are likely to be met.

4.1 Women as Entrepreneurs

This section is divided into two parts. It starts with a brief description of the concept of entrepreneurship and an overview of the gender gaps in entrepreneurship which is followed by an overview of the general characteristics of women's enterprises, focusing on the Global South, and several of the challenges they face as entrepreneurs, although many of which are universal.

There is no universally agreed definition of an 'entrepreneur'. The term is often used to imply someone who is self-employment (sole proprietorships versus employer firms). However, in the academic literature many authors do make a distinction between someone engaged in running a business, and someone who is innovative in the sense that they offer 'products or services that are *new* [authors' emphasis] to some or all customers, with additional consideration for the extent that no or few other businesses offer that same product' (Kelley et al. 2013: 32; Shankar et al. 2019). Given that this chapter focuses on providing energy services and promoting energy access which are more than likely to be based on new technologies or energy sources in the setting where they are being produced or sold, this focus fits well with Kelley and her colleagues' definition, so it will be the one used here. By extension, an entrepreneur who is innovative is exhibiting entrepreneurship. There are gender differences in all aspects of entrepreneurship. For example, evidence points to self-employment at start-up as a more common route to income generation for women than men (Johns Hopkins University et al. 2019) with women globally, more likely than men, reporting job scarcity as the motivation to start up a business (GEM 2021). Age is also a factor in taking the decision to start an enterprise with both women and men entrepreneurs primarily in the age range 18 to 34 initiating a start-up, linked to members of this age group being less likely to have monetary responsibilities, whereas education is not so strong a determinant (GEM 2023b). Women-led businesses are considered to focus on social benefits, which would link with promoting energy access, and appear to be more likely to hire women. Women's businesses tend to be smaller in terms of employee numbers. While start-ups are often initially solo efforts, more women than men stay as sole proprietors which, as is shown below, is not necessarily because they lack ambition to grow. There are regional variations, with Italy having the highest percentage of women sole proprietors (83.3%) and the United Arab Emirates reportedly the lowest at 2.6% (GEM 2021). The Global Entrepreneurship Monitor (GEM), which annually surveys global entrepreneurial activity in around 60 countries with good regional spread, finds that women and men agree that being an entrepreneur is a good job with high-status.

When establishing a business, an entrepreneur will face a number of challenges which can prove more difficult for women to overcome than their male counterparts. First, in some cultures, there are social norms which are not supportive of women working in general. Another way in which social norms influence decisions about work relates to the nature of what is deemed an appropriate occupation for women and men with gender stereotypes playing a role (IDS and GIZ 2019). For example, manufacturing and construction are considered to require physical strength, which is regarded as a male characteristic, and therefore not appropriate occupations for women and hence women play only minor roles as entrepreneurs in these sectors (GEM 2012). This is potentially a significant finding if we expect women as energy entrepreneurs to be involved along the value chain, including as producers of equipment rather than only as sales agents.

Having taken the decision to establish a business or upgrading an existing enterprise, for example, accessing modern energy services, there are practical issues such as lack of access to finance, often linked to the need for tangible assets as a guarantee, such as owning a house or land, which the regulatory system, such as inheritance laws, can be biased favourably towards men. However, evidence shows that compared to men, women face greater difficulties in accessing capital to invest in their enterprises (Cecelski 2005; Nordman and Vaillant 2014; Baruah 2015; Chopra 2015). Barriers for women entrepreneurs to access finance is a global issue. Data for Europe shows that the capital invested in European technology companies 91% went to all-male founding teams while 7% went to mixed teams and less than 2% to all-female teams (European Commission 2022). This reluctance to support women-led companies might appear surprising since they require less capital at start-up than the all-male and mixed teams while delivering higher returns (Fackelmann and De Concini 2019). Women entrepreneurs also face prejudice from financing agencies who consider that women lack motivation to grow their businesses, that their enterprises are survivalist, established out of necessity, not likely to generate profits and so making repayment of loans difficult, therefore based on these criteria women entrepreneurs are evaluated as not a good financial risk (Berner et al. 2012). Men, on the other hand, are seen as ambitious and growth-oriented (University of Twente et al. 2015). However, research in five sub-Saharan African countries found that women entrepreneurs did not see themselves as survivalist but aspired to grow the same as their male counterparts (ENERGIA 2019). Women can overcome the challenges of accessing finance from formal institutions by using Community and Savings Groups (Dutta et al. 2017) as well as micro-finance institutions[1] (MFI) which generally consider women to be less of a credit risk than men based on perceptions of women tending to value access to credit since it is and that they avoid wilful non-payment (Arunachalam 2008).

Many women lack not only the appropriate education and skills for running a business, such as basic arithmetic and literacy, but also some basic understanding of science and practical skills, such as metal working, which would be relevant for the energy sector. Women's literacy and numeracy levels can also be problematic acting as a barrier in applying for training and support, as well as making women reluctant to apply for loans. Several trades offer apprenticeship schemes which are a route to gaining the skills and experience to open and run a business. Evidence from Ghana indicates that in trades dominated by men, such as car maintenance, apprenticeship programmes appear to be longer and more expensive than those in typically female trades such as tailoring or hairdressing (IDS and GIZ 2019). The implication of the differential in costs is that young women may be deterred by their family from entering trades with potentially higher levels of remuneration. In households with limited incomes, families may decide it is better to invest in a potentially higher income earned by the young men of the household than investing in young women's careers. Nevertheless, women do defy gender norms and enter trades usually considered male preserves as has been happening in the Global North for some time. Evidence shows that these changes are also taking place in the Global South, for example, young women in Ghana have become motor mechanics and young men have taken up hairdressing which seems to create a novelty value and so generates customers

[1] A micro-finance institute provides a range of financial services to individuals, entrepreneurs and small businesses who do not have access to traditional banking.

(IDS and GIZ 2019). Male role models and family support to try techniques used in manual trades are found to help young women cross the gender divide in employment (IDS and GIZ 2019).

In many cultures women are assigned the responsibilities of care for children, the sick and older family members which influence women's approach to entrepreneurship in several ways. Firstly, the double workload of combining running a business and care tasks can be a barrier to establish or continuing a business (Jennings and McDougald 2007). A strategy women adopt to balance the workload is to run their business from home. Another reason for working from home relates to safety since anecdotal evidence reports harassment of women operating businesses in public spaces (Vossenberg 2013). However, evidence points to negative perceptions by potential customers and financing agencies about businesses operating from the entrepreneur's home (Marlow 2002 cited in Vossenberg 2013: 5). Home-based business may also miss out on potential government financial support, for example to promote energy efficiency, because government agencies can overlook these productive activities since they can be indistinguishable from other household activities, such as food preparation.

Not surprisingly, the global health pandemic which took hold in 2020 has had an impact on businesses. The GEM surveys for 2021 reports that, globally, more women than men reported closing a business due to the pandemic (41.9% vs. 35.5%) although there are regional variations (GEM 2021). On the other hand, the pandemic also provided business opportunities, with men more likely than women to report start-ups with the gap most pronounced in high-income countries. Women entrepreneurs in low-income countries were particularly satisfied with their government's responses for business support during the pandemic.

Many women's businesses, at least at start up, would be classified as micro- and small-enterprises (MSEs),[2] many of which are in the informal sector[3] Working in the informal sector has been driven by urbanisation

[2] There is no universal agreement on what constitutes a micro- or small-enterprise. Data from the International Finance Corporation (IFC) indicates that the number of employees is a common measure for classification with micro-enterprises falling in the range of 0–9 employees, small businesses in the range of 10–49 employees, medium-sized businesses 50–249 employees and large businesses 250 employees or more (Haider et al. 2019).

[3] There is no universal agreement on what constitutes 'the informal sector or informal economy'. In part this lack of definition can be attributed to its neglect by development

and economic crises (University of Twente et al. 2015). While accurate data, particularly sex-disaggregated data, about the informal sector is difficult to find the indications are that the informal sector plays a significant, but often unrecognised, role in national economies. A World Bank study in Senegal estimated that the informal sector in 2007 contributed 12% of the country's GDP (World Bank 2007). In the Southern African region, after agriculture, the informal economy is the most important source of employment, providing 72% of jobs (Charmes 2012). In India, it is estimated that the informal sector contributes around 50% of the gross value added to the economy (Sanjay et al. 2023). For informal sector enterprises operating outside of the household, urban planners have negative views about them seeing them as dirty, dangerous, disorganised, backward and undesirable. Such negative stereotyping feeds into and is feed by theories of modernisation which have underpinned approaches to development in which governments, supported by international agencies, aim to formalise activities in all sectors. Rwanda is an example of a country taking this line (Mohlakoana et al. 2018). There are justifiable concerns that workers have limited benefits beyond direct earnings, for example, they are not covered by any labour protection laws or may have restricted access to public services (Ishengoma and Kappel 2006). On the other hand, informal sector enterprises exist because it is relatively easy to set up an enterprise which draws on your existing capabilities (hence why so many women's enterprises are based on housekeeping skills such as cooking), requires a minimum of capital investment and can rely on local resources for inputs. An aspect of informality is that entrepreneurs may sell their products in town or city centres in informal structures based at a fixed or semi-fixed location while others are ambulatory using carts

economists until an economic anthropologist working in Ghana pointed out the informal sector's significance for the local economy (Hart 1973). Since then, a lot of the effort by international agencies, such as the ILO, has gone into making a definition that can allow for the development of measurement criteria. In the context of this book, I have taken the informal sector to be broadly characterised as comprising of enterprises that produce goods and services for sale which operate on a small scale with a low level of organisation involving only a small number of people. If an informal sector enterprise involves more than the owner, then the people who work there are employed on a casual basis, most probably with a kinship or personal and social relationship with the owner. The term 'informality' applies because the enterprise does not comply with a state's legal requirements for registration. These requirements vary from country to country but are intended to ensure businesses comply with national legislation across a variety of policy tools, for example, tax or social security obligations or environmental standards.

or head-loading. Municipalities have a range of reasons for removing urban street traders, including health concerns related to food sold from stalls without access to clean water and rubbish disposal. For example, the Department of Health in the Philippines found that seven out of ten roadside food handlers to be infected with parasites (Lumampao et al. 2006). There are solutions to these problems which simultaneously meet the goals of municipality officials and informal sector workers. A more effective approach might be to work with the people whose livelihoods are affected by regulations. An example comes from Dar es Salaam, Tanzania, where a project worked with street traders to design the stalls that would allow them to display their wares which generated a positive feeling in the traders (University of Twente et al. 2015).

MSEs are often based in or operate out of the household producing, processing and repair, as well as selling a wide range of goods and services. For example, in rural India and Nepal, typical household industries and enterprises include foodstuffs (production of flour, dehusking of paddy, grinding of herbs, preparation and sale of pickles and milk products), beverages (manufacture of liquor, soda water, etc.), textiles (weaving of cotton, jute, silk) products made from wood, paper and leather and chemical products (paints, colours, inks, etc.). Social changes, in particular male migration to urban areas, have led to more women being involved in running enterprises (University of Twente et al. 2015). Women's enterprises are generally located in low-growth sectors, often linked to domestic activities such as cooking and laundry, which are characterised as having low productivity and low profit, with limited use of modern technology (de Haan 2016; Cirera and Qasim 2014; Klapper and Parker 2011; Dejene 2007; Hunt and Samman 2016).

An enterprise can be positioned at any point along a supply chain which can be gendered. For example, in Nigeria, men dominate grating, dewatering, milling operations in cassava processing, whereas women have responsibility for peeling, washing, drying and frying (Davies et al. 2008 cited in University of Twente et al. 2015: 20). However, care needs to be taken in treating women and men as homogeneous groups when assessing differentiation in tasks—differentiation can also occur within groups of women. A simple gender disaggregation of milk and dairy chains in West African would show that it is men's responsibility for the care of the livestock but the milking and dairy processing is women's responsibility—however a deeper analysis shows that this chain is governed by a complex set of social arrangements based on ethnicity and religion. For example, in Nigeria and Senegal, Fulani women process the household's milk and make decisions about how much to send to market. However, wealthy Fulani women and women conforming to a strict Muslim code often use women from other groups as intermediaries (Waters-Bayer 1985; Corniaux 2003).

Being part of an association can provide advantages for an entrepreneur. An association can have different forms of organisation. For example, women have established cooperatives in the dairy chain in Kenya, Senegal and Rwanda, which enable them to negotiate contracts and fairer prices, as well as being able to speak with a united voice with the authorities (University of Twente et al. 2015). Similar benefits were reported that membership of the male-dominated Association of Small-Scale Industry in a Ghanaian town included: 'pooling resources for specialised training, engaging with the authorities to get fair tax bills, organising apprenticeships, and targeting markets together—for example, by persuading local institutions to buy local' (IDS and GIZ 2019: 45). A study of households involved in handloom production in the state of Andhra Pradesh, India, indicates that energy may not be the only enabling factor in improving productivity. Households which received non-energy inputs, such as training, access to good quality raw material and market access via a formal institution had a significantly larger increase in household income than those households that were not linked to any such formal institution. (47% of households in the supported communities report an increase in income with only 6% of those in those communities without institutional support.) (Garimella et al. 2015).

4.2 Energy Enabling Women's Income Generation

Women's micro-enterprises are often heat-intensive (food processing), labour-intensive, and/or light-intensive (home-based cottage industries working in the evening) with implications for women's health. Indeed, there appears to be little research into the health impacts of exposure to local-level pollution in informal sector enterprises. The use of unprocessed biofuels and residual mineral oil products to generate process heat is common. So, a reasonable assumption is that the effects of exposure for enterprises not based in households would be similar to those from kitchen wood stoves. However, to heat larger quantities of materials, entrepreneurs can use discarded oil drums as make-shift combustion devices which may expose operators to higher levels of pollution. For example, fish smoking in urban communities in Lagos, Nigeria, can involve women standing for seven hours a day smoking fish using fuel-wood in old oil drums exposing them not only to the wood smoke but also other pollutants from residual oil of unknown toxicity (Maduka 2006). Informal sector enterprises are often located along the side of roads resulting in the entrepreneur's exposure to vehicle exhaust fumes. To improve working conditions and productivity requires investment in equipment, and particularly for women, equipment that would free time related to household tasks. However, as was pointed out in Sect. 4.1, women face higher barriers than men to access finance needed to buy equipment because of their lack of ownership or control over resources needed to underwrite any loans.

Food processing chains involving both women and men use metabolic energy for transportation and many of the processing tasks. However, it appears that for food processing chains, with the possible exceptions of milk churning and cheese making, energy access for income generation focuses on cooking (thermal) energy neglecting heat for drying and mechanised power for the other processes and transport (University of Twente et al. 2015). Reducing arduous physical labour through mechanisation can contribute to reducing strains and accidents which may have an impact on a person's capacity to work with the knock-on effects on income. As was pointed out in Chapter 3, incapacity to work is considered the most overwhelming reason why households become poor (CPRC 2008).

Evidence points towards men's enterprises being more likely than women's enterprises to use modern energy carriers, particularly electricity. One of the few studies of the productive uses of energy from a gender perspective, found when assessing the impact of the access by MSEs to modern energy, in the three focus countries,[4] that if interventions promote electricity as the energy carrier men entrepreneurs are the main beneficiaries (IDS and GIZ 2019). Evidence shows that the business outcomes of both women and men entrepreneurs using electricity are better than for other energy carriers. A three-country study which assessed the influence electricity had on women's income found in Kenya that women reported their income doubled while in Nepal and India women reported that electricity access made little difference to their

[4] Ghana, Tanzania, Myanmar.

income (University of Oslo et al. 2019). This finding suggests that there are more factors influencing enterprise success than energy access.

Women's enterprises tend to have technologies with lower efficiencies and productivity compared to their male counterparts in like-for-like enterprises. The Mali Multifunctional Platform (MFP) programme[5] focused on improving equipment efficiency which resulted in increased incomes (Sovacool et al. 2013). Other ways that incomes improve through energy access, include being able to relocate your business to reduce costs. For example, a tailor could work from home using a solar light meaning that there was no longer the need to rent a room with electricity (University of Twente et al. 2015).

Electricity allows for diversification in the services offered by enterprises, for example, a range of food stuffs and drinks. A guest house owner being able to pump rather than head-loading water saves the owner time as well as providing a better quality of service for guests—which improves income (IDS and GIZ 2019). However, electricity is not the most cost-effective option for process heat which is the energy form used by many women's enterprises. Also, small stand-alone systems, such as SHSs, do not produce the level of power output suitable for driving some machinery for which grid electricity is a better option provided that the supply is reliable! This unreliability, which can damage equipment, has opened the opportunity for private operators to set-up mini-grids offering higher-quality service, with better availability than the main grid. However, the cost per kWh may be significantly higher than the main grid (IDS and GIZ 2019).

One of the few pieces of evidence related to energy access and women's enterprises in the Global North relates to the lack of protection provided to small businesses, such as hairdressers or coffee shops, against rising energy prices and disconnection in the event of non-payment of bills. It is estimated that within the EU, more than 60% of employees in service, sales and clerical support are women who will be in part-time, temporary,

[5] This programme, running from 1999 to 2004, was a development assistance initiative in rural Mali related to improving energy access that distributed small diesel engines attached to a variety of end-use equipment (hence the name multifunctional platform) which provided villagers with income-generating projects through improved energy availability. The Mali government, which had been involved from the start of the programme, installed another 1800 MFPs by the end of 2012 (Sovacool et al. 2013).

low-paid and precarious employment often in SMEs (European Commission 2022). Since these businesses are often significant energy users, their viability is vulnerable to energy prices and are often located in rented premises dependent on the willingness of the landlord to improve the energy efficiency of the building. A welcome development in the European Union is the establishment of a Social Climate Fund which has introduced the concept of 'vulnerable micro-enterprises' which enables the provision of support for SMEs including those services industries where women work (Clancy et al. 2022).

4.3 Women Working in the Energy Sector

Woman are underrepresented in the global energy sector workforce. Figure 4.1 shows the percentage of women in the workforce of a range of industries from 2016 to 2020. Sectors which have more direct contact with clients, such as health care, have a higher percentage of women employees (Froehlicher et al. 2021). Although it is possible that the level of women's employment is underestimated due to the way data are compiled. For example, a study in Pakistan of the brick kiln industry found that approximately 100,000 women worked together with other family members in the industry, but women were not classified as officially employed since only men, as head of the households, were registered (quoted in Kelkar and Nathan 2005). Yet, the International Renewable Energy Agency (IRENA) warns of a skills shortage across the renewable energy value chain which is increasing in importance as part of the energy transition (IRENA 2013). Arguments for increasing women's participation in the energy sector come from three perspectives. Firstly, women's own aspirations to have an interesting, socially relevant and well-paid job or contributions to women's economic empowerment, gender equality and challenging patriarchy working in the energy sector.

Secondly, for energy sector organisations, women employees can be considered good for business. Empirical evidence shows that having a diverse workforce improves not only a company's financial performance but also its reputation (EY 2014).

Thirdly, from the government's perspective women's participation in the energy system contributes to economic and social goals (GEM 2023a). Evidence indicates that at the macro level the economy suffers by not making use of women's skills and talents (World Bank 2011). Given the key role energy plays in the economy, it is not unreasonable to assume that the energy sector would be influenced by women's participation. Encouraging women to work in the energy sector promotes gender equality and in so doing contributes to reaching SDGs 5 and 7.

This section is divided into two parts reflecting the ways in which women can work in the energy sector: as entrepreneurs running energy service companies and as employees in utilities and large private sector energy companies.

4.3.1 *Women as Energy Entrepreneurs*

The arguments for women becoming energy entrepreneurs are varied depending on the perspective. There is the gender equality argument and there is an instrumental argument that women have a key role to play in promoting the energy transition since they have better access to women customers and potentially employees. While there are claims that women's engagement in the energy sector contributes to women's empowerment, it should be kept in mind that energy organisations promoting the energy transition and energy access are not principally focusing on women's empowerment but on providing services which generate a profit. Unfortunately, the empirical evidence related to the outcomes of women as energy entrepreneurs is weak. A comprehensive review of the literature related to women's energy entrepreneurship found that the available evidence consists primarily of qualitative, anecdotal and descriptive studies and programme reports much of which is in the grey literature (Johns Hopkins University et al. 2019). There appears to be little empirical research that explores the counterfactual position that having female energy entrepreneurs leads to greater or reduced energy access compared with the situation of not involving women.

This sub-section has three objectives: (i) it summarises the evidence relating to women as energy entrepreneurs; (ii) it reviews critically the emerging business models to evaluate the impact on transforming gender roles and relations; and (iii) it documents the barriers to women's participation as energy entrepreneurs.

The appearance of advocacy for women as managers and entrepreneurs of energy services and products could be credited to Batliwala and Reddy (2003) who argued, based on evidence from the SEWA (Self-employed Women's Association) in India and the Grameen Bank in Bangladesh, that energy projects should move on from seeing women only as beneficiaries to a more active role in delivering energy access. Several arguments are presented by development practice for women as energy entrepreneurs in rural areas. Women are usually their household's energy manager so will have a good understanding of potential customers' needs. Cultural norms can make it easier for a woman than a man to enter the household of a woman who is not a family member, as well as developing trusting relationships which form the basis of commercial transactions.

One of the few studies of energy entrepreneurs which takes a gender approach, finds a difference in motivation between women and men to become energy entrepreneurs. The study is a qualitative analysis of Solar Sister,[6] a social enterprise based in East Africa, which sells solar lanterns and clean cookstoves through a supply chain of women and men entrepreneurs (Soria et al. 2016). Women who became entrepreneurs reported that the main motivation was to have their own source of income to support their families, as well as having access to products that saved on energy costs. Seeing the success of other women as Solar Sister entrepreneurs was an additional motivating factor for some women. For the men being a Solar Sister entrepreneur was not their primary income-generating activity but provided either a secondary source for farmers, generating income more evenly spread through the year, or for shopkeepers by diversifying the products they could offer. In another study in Rwanda of women and men operating village-level enterprises (VLEs) supplying rechargeable solar LED lights, a small percentage of women reported that the social interaction they enjoyed as part of their involvement in VLEs was part of the motivation for their participation (EPRU (UCT) and IPA 2019).

It is also possible that becoming an energy entrepreneur may not always be a woman's primary income source. Women based in rural areas who become energy entrepreneurs may see these activities as supplementary to income from agricultural activities (Pailman 2016). This business model has consequences for the client base, particularly for those selling energy products rather than services based on modern energy. Firstly, if the service operates on a seasonal basis products that the women are selling may not always be available as needed. Also, if expanding their business requires finding new clients outside of the immediate local and social network, travel is potentially going to be more difficult if attention to agricultural activities takes priority.

[6] https://solarsister.org/about-us/our-origin/ (accessed 25 July 2022).

Women's engagement in the energy sector takes place in a range of organisational forms: community self-help groups, cooperatives and MSM enterprises (which can be in the formal and informal sectors). Self-help groups (SHGs) are formed by a voluntary association of individuals at the village-level meeting regularly to undertake an agreed activity using their own resources (such as labour and finance—savings and loans associations are often SHGs). SHGs have been found to promote entrepreneurism (Satyasai et al. 2014 cited in Johns Hopkins University et al. 2019: 43). SHG involvement in energy entrepreneurism involves selling energy products door-to-door and from stalls in local markets. SHGs can also act as distributional channels for energy technologies. For example, in Bihar where a project, as part of an SDG initiative,[7] would only allow women members of SHGs to purchase, with the help of a loan, the solar home systems (SHS) it was promoting (University of Oslo et al. 2019). Also, in India, SHGs have set up machine-hire centres which enabled women to have access to farming equipment that they would find difficult to buy as individual items (MSSRF and CRT Nepal 2019). While this model of enabling access for many low-income households, often in remote, difficult to access locations, has had considerable successes in terms of numbers reached, there is little evidence available as to whether women's SHGs move beyond acting as distribution points.

In contrast to SHGs, there are examples of village-level cooperatives established as local energy enterprises involving women. Some cooperatives are producers of energy technologies, mainly linked to improved cookstoves, as well as distributors of equipment. Membership of a cooperative provides opportunities for women to learn technical skills as well as basic literacy and numeracy skills. An early example of a women's energy cooperative comes from the remote island of Char Montaz in the south of Bangladesh. The cooperative was established in 1999, with World Bank assistance, to produce fluorescent battery-powered lamps to replace the traditional kerosene lamps in a location where grid electricity is a distant reality. The lamps would also improve the lighting and indoor air quality in rural homes as well as reducing fire risks. When the project started, if a woman constructed and sold two lamps a day, she could earn a daily wage equivalent to a skilled labourer. The project design took the local context into account, first, in recognising women would need technical

[7] Lighting a Billion Lives https://www.gogla.org/lighting-a-billion-lives (accessed 1 June 2022).

training as well as business skills, such as accounting and book-keeping. Secondly, to overcome any potential male resistance, male family members were encouraged to act as business advisors to the women (Khan 2001).

Perhaps the most famous energy 'co-operative' is *Windfang*, which operates grid-connected wind turbines and solar plants in North Germany (Delfs 2000). The company was founded in 1991 by three women with technical backgrounds who had grown disillusioned about working in an unsupportive male-dominated energy sector. Strictly speaking it is not a cooperative, but a company owned by a group of women which works on cooperative principles using a democratic structure. Not all cooperative members work in the company; some women provide financial support seeing this as a contribution to the energy transition. All members elect the steering committee and the board as well as being involved in major decisions taken at a general assembly. Windfang provides a model which actively promotes the inclusion of women irrespective of an individual's skills. The membership is multidisciplinary from a broad range of disciplines, including theatre and economics, and with varying degrees of experience from students to women professionals representing all age groups. An important aspect of the way Windfang works is the emphasis on providing a supportive environment to women to acquire the skills to work in the energy sector. A measure of success of this approach is that women have left the cooperative and established their own businesses or found employment in the expanding renewable energy industry.

Another example, of how solidarity in the workspace can contribute to energy entrepreneurship comes from the Solar Mama Programme (SMP) of the Barefoot College in Rajasthan. The SMP has two programmes: one covers the building, operation and maintenance of small-scale energy photovoltaic (PV) home systems and solar lanterns while another offers a similar programme for solar cookers. There is also an international programme in which women from all over the world are allowed to participate in both programmes. The training is residentially based and aspiring solar engineers, who are illiterate and semiliterate rural women from across India, live on a campus for six months (Mininni 2019). Two women from unelectrified villages are selected by the College, with community inputs, as trainees. The College makes an agreement with the village that the elders of the community will form a Village Energy and Environment Committee (VEEC), membership of which comprises of both women and men, to collect funds from each participating household to support the solar programme and to provide a workshop space where the engineers can work on their return (Barefoot College 2015 cited in Mininni 2019). The College is influenced by a Gandhian egalitarian philosophy and operates in a no-hierarchical way. After completing the training programme, in 2003, a group of engineers established the Women Barefoot Solar Cooker Engineers' Society (WBSCES) which, in India, possibly represents the first society set up by illiterate and semiliterate rural women. Within the Society, the women have added entrepreneurial skills, such as accountancy, management, purchasing, sales and budgeting, to their technical competences (Kakani 2014 cited in Mininni 2022: 118). The society also provided a space in which women could support each other to challenge existing social structures that act as a barrier to women's participation in the solar value chain. In Rajasthan villages, Muslim women are not allowed to leave their homes, but the WBSCES was able to persuade some families that they provided women with a safe place to work. Another cultural norm WBSECES challenged was the unacceptability of women approaching local (male) traders with initially the traders refusing to deal with them. The nature of some activities related to building solar cookers challenged the caste system that dictates roles and status (Sahgal et al. 2021). Welding and working with iron are activities assigned to low caste men, but when women do this work, they challenge not only gender norms but also the caste system.

Organisations that are part of a supply chain can also provide a space for their agents to enhance business skills, build networks, develop friendships and share learning experiences. An example of this approach is from Solar Sister which holds regular 'sisterhood' meetings for the women involved in their chain.

The lack of technical skills can be a significant barrier to women becoming energy entrepreneurs. A project in Zambia to establish solar home system installation and maintenance selected interested entrepreneurs for training if they had knowledge of electricity and electrical systems (Munyeme 1999). This criterion ruled out most women. On the one hand, it is a positive development that organisations have learned from these early experiences and there are initiatives to train women with the skills to work in the sector, in the Global South primarily driven by development assistance funding. On the other hand, these initiatives do not always live up to expectations in terms of their outcomes. An independent analysis of a Grameen Shakti project in rural Bangladesh aimed to increase the number of women with the technical skills to assemble, mount and maintain solar home systems found the long-term results disappointing (Hemson and Peek 2017). From 2005 to 2010, the company trained 2797 rural women. At the time of the study survey in 2013, it appears that Grameen Shakti was employing in their value chain less than 3% of the women that they had trained. Nevertheless, some of the project trainees had become involved in income-generating activities outside of the energy sector. The women working for Grameen Shakti were not using all their acquired skills of promotion, installation and maintenance of the SHS but were only used to assemble SHS components. Indeed, the SHS value chain (production, assembly, marketing, sales, distribution, installation and servicing) appears highly gendered with women dominating assembly and men installation. An explanation for this chain construction lies in gender norms, what is considered right and proper for women and men to do. Around a quarter of the women engineers and nearly a third of regional managers interviewed felt that SHS installation work was too heavy for women.[8] The opinions of managers (who, at senior level, are all men) were divided as to whether it was appropriate for women to travel to villages. The project's aims included trainees becoming entrepreneurs within the SHS value chain. However, Hemson and Peek (2017) could find no evidence of women acting as SHS entrepreneurs.

Gender norms about what constitutes appropriate employment for women and men have been found to be influencing career choices in the Netherlands (Feenstra et al. 2022). Employment as an engineer was considered dirty and dangerous work, possibly based away from home, whereas employment in the IT field was perceived as working in an air-conditioned office, smartly dressed, with a daily commute from home. The latter was considered a more appropriate career option for women. A programme to promote access to low-cost solar water heating solutions and solar dryers in Georgia, took a gender-sensitive approach to job creation aiming for equal involvement of women and men. Participants could opt for training in construction or maintenance/monitoring which resulted in women opting for the latter and men the former because construction is culturally perceived as a man's job. However, women found themselves curtailed in their chosen field by maintenance requiring them to travel alone which was also time-consuming and did not fit with their family responsibilities (WECF 2015 cited in Glemarec et al. 2016).

[8] This is in agreement with the findings of Bose et al. (2009) who looked at women's participation in employment across a range of sectors in rural Bangladesh.

Also, gender norms can act as barriers to training. A stove project in the state of Karnataka, India, required women and men to travel outside the state for their training. However, in some villages, comments about the morality of women who undertook such a venture, made women reluctant to participate in the project (Govindan et al. 2019). Also , the appropriateness of clothing can play a part in women's involvement. In a grid extension and maintenance programme in the state of Maharashtra, India, women are given uniforms to enable them to deal with live wires and climb poles. They are similar in design to those of men with the idea that this would promote a feeling in the women of equivalence with their male colleagues. However, the women were reluctant to wear the uniform since it did not resemble their traditional clothing (Govindan et al. 2019).

Sometimes existing norms can favour women's inclusion in the chain, for example, in areas where programmes are promoting clean cooking solutions and norms do not allow men who are not family members to enter a kitchen—this creates opportunities for women to be involved. Norms vary with context. A biogas programme in several countries in Sub-Saharan Africa, which targeted women to take up non-traditional roles as masons to build digesters gave, as of 2012, the percentage of women masons as: Kenya 12%, Tanzania 5%, Ethiopia less than 1%, Burkina Faso 1% and Uganda 6% (InforWit Research Consultants 2013). It is possible that in the biogas programme another cultural normally is working positively in women's favour, since in Kenya women are considered neater workers than men—so customers preferred women to do the job building digesters. The limited independent evidence about women in marketing, sales and distribution of energy technologies found that they could outperform men, because they were trusted more than men to provide quality service (University of Oslo et al. 2019). On the other hand, women had practical arguments for not wanting to be a mason. Firstly, carrying heavy loads of bricks is demanding of metabolic energy, which women felt does not leave much energy for them to fulfil their family responsibilities. These responsibilities were also used to explain why some women did not want to be away from home for training or to build digesters. Usually, women will manage their time to combine their family tasks with any farm work which tends to be near their house (Clancy et al. 2016).

In the Global South, a survey of the ten finalists for the 2014 Ashden Awards[9] found that the average rate of direct female employment (not including entrepreneurs or sales agents) was 23% (Wheldon et al. 2014). The same survey found that women's participation along the energy value chain is generally low unless women are specifically targeted for inclusion. So where along the energy value chain are women entrepreneurs located? Most of the available evidence points to women being in marketing, sales, distribution and management of services related to energy technologies. To reach low-income consumers with lighting and cooking solutions, a growing number of energy enterprises have begun to employ women as sales representatives. Such a strategy assumes that women will help ensure that energy products reflect the priorities of women users, thereby increasing the likelihood of adoption and use. Being in aftersales services of maintenance and repair can offer an entry point to women with limited access to capital since there is likely to be a lower investment required for start-up, provided they can find training in the appropriate technical skills. Another reason for employing female technicians to repair equipment is that in some cultures, it is inappropriate for male technicians to enter the homes of customers while male family members were away (Baruah 2016). These strategies are not specifically designed to empower women but part of a business plan to generate profit.

There is very limited evidence about women entrepreneurs directly involved in design and development of energy technologies. A frequently cited example is the Upesi Stove project in Kenya in which women were involved in design and field testing of stoves (Khamati Njenga 2001). Women entrepreneurs can be involved in design indirectly. For example, women entrepreneurs in Tanzania who were primarily involved in operating solar-powered kiosks for mobile phone charging provided feedback into the design of the kiosks the nature of which could help encourage more clients (Pailman 2016).

[9] The Ashden Awards are annually awarded to organisations promoting renewable energy innovation both in the UK and the Global South. https://ashden.org/ (accessed 2 February 2024).

4.3.2 *Women Employed in the Energy Sector*

This sub-section examines the employment patterns for women in the energy sector. The data are rather limited despite the efforts of the International Renewable Energy Agency (IRENA) to obtain a comprehensive picture through its surveys (IRENA 2019).

The available data show that globally the energy sector is typical of any sector dominated by technology, it is primarily a male preserve. For example, in the European Union, the average is, as of 2010, men at 77.9% and women 22.1% of the work force in the energy sector (EIGE 2012).

Engineers are usually men with women generally found in non-technical, administrative positions and public relations with limited decision-making power and lower salary levels (Pearl-Martinez 2014). The occupational trades, such as metal workers, insulations specialists, plumbers and pipe-fitters, electricians, heating and cooling experts, are similar in their gender balance to the professional posts, in other words they are mostly male-dominated (Sustainlabour 2013). The World Petroleum Council estimated that, as of 2017, women made up 22% in the global workforce which is lower than the average in the total workforce. Very few women reach senior positions. Data from 2016, shows that women made up 16% of board members of the global top (by revenue) 200 utilities.[10] In the oil-and-gas sector women make up around 27% of graduate entrees while it is estimated that only one in a hundred CEOs is a woman (Rick et al. 2017 cited in IRENA 2019: 29). The renewable energy sector appears to be only slightly better than the oil and gas sector (EIGE 2012). A survey of 90 companies involved in the renewable energy value chain gives an estimate of women making up approximately 35% of the workforce (IRENA 2019). Women tend to dominate part-time working, for example, in the renewable energy sector in Spain, in 2006, 2% of jobs were designated part-time of which 67% were filled by women (Arregui et al. 2010 cited in Baruah 2017).

A survey of the US solar industry shows a trend of increasing female participation but with continuing discrimination for women and men of colour—both earn less than their white counterparts in comparable positions (Solar Foundation 2018 cited in IRENA 2019: 30). If this pay differential leads to fewer people of colour being part of the industry,

[10] http://www.ey.com/gl/en/industries/power---utilities/women-power-and-utilities (accessed 15 November 2016).

there is evidence to suggest that this will be detrimental to the organisation. A diverse workforce provides a range of different perspectives and understanding about a situation, such as the energy transition, and evidence shows that diversity improves financial performance (EY 2014). Canada has been among global leaders in addressing employment diversity issues. The oil and gas sector aims to create a diverse workforce using policies and management practices that encourage representation and full contribution of individuals from all societal groups. A significant aspect of the strategy is that recognition is given to staff contributions on the basis that they are capable individuals, rather than as members of designated groups which is an approach intended to overcome any resistance to what is seen as privileging one group over another (Dowse et al. 1999).

Evidence suggests that women establish themselves more easily in newly emerging sectors, rather than long-established ones (Pearl-Martinez 2014). So why does a gender gap continue in the renewable energy sector? Based on a review of the evidence, Clancy and Feenstra (2019) identified a range of explanations:

- lack of appropriate skills due to the gender gaps in energy-related education;
- the perception of the energy sector as a male domain and persisting gender stereotypes;
- the difficulty of achieving a work–family life balance which discourages women from taking on jobs that involve unpredictable work schedule or emergency travel;
- insufficient career promotion opportunities and mentoring programmes for women.

4.3.2.1 Gender Gaps in Qualified Personnel

The number of women graduating with a degree in engineering or technology is significantly less than the number of men which is reflected in the levels of employment. In the United Kingdom, between 2015 and 2016, only 16% of women entrants into the workforce had an engineering or technology degree while women make up nearly half of the total workforce (47%) (Engineering UK 2018). Data for the apprentice level in the UK show, between 2002 and 2014, an unexplained decline in female apprentices in engineering from 4.6% to 3.8% (Young Women's Trust 2016). On the positive side, the number of women scientists and engineers within the EU appears to have increased. There were nearly 6.6 million women working as scientists and engineers in 2020, which is 254,500 more than the previous year (Eurostat 2022).

Engineering as a career appears not to be attractive to young women. A quarter of girls aged 16–19 who took part in a UK survey said they would never consider a career in engineering (Engineering UK 2018) with similar reactions across several EU Member States (Belgium, Finland, Poland, Spain and Sweden) (OECD 2014). In part this reflects the images young people have about the nature of engineering. Secondary school students in Canada said that they associate engineering with 'construction work, outdoor work, working in a cubicle and relating primarily to computers and machines rather than to people' (Baruah 2017: 21).

Gender stereotyping influences women and men about what in a particular society is considered an appropriate activity, including types of employment, for women and men. There is evidence to suggest that children have internalised gender stereotypes by the age of four (UNESCO 2017). Further reinforcement comes through socialisation in which girls are encouraged to be social and engage in group activities, whereas boys are encouraged to explore the physical world as individuals, discover how things work and solve problems. When it comes to careers and choosing appropriate subjects to meet those aspirations, family and friends play important influencing roles. Mothers, in particular, are recognised as having a strong influence over their daughters about career choices (Harackiewicz et al. 2012). However, the evidence is contradictory about the outcomes in respect of choosing science, technology, engineering and mathematics (STEM) subjects. Some researchers find that mothers positively motivate their daughters to persevere with science and math (Leaper et al. 2012). Whereas other studies find that, at least in the early formative years, mothers are more likely than fathers to re-enforce the notion of the masculinity of STEM subjects hence discouraging girls from studying them (Yee and Eccles 1992; UNESCO 2017). Girls tend to opt for careers which are considered to have strong social goals (such as social work, nursing, teaching and human resources) and reject careers seen as dissociated from social goals (such as astronomy and engineering) (Su et al. 2009; Diekman et al. 2010). A global survey of the reading, mathematics and science capacities of 15-year-olds found that, compared to boys, girls have a low self-belief in their innate ability to achieve goals in science.[11] Surprisingly the gender gap was found to be particularly large in Denmark, Iceland and Sweden, countries generally associated with good gender equality (OECD 2015).

Once children enter their teenage years, peer-groups become a major influence which is when many young people begin to make decisions about careers. Indeed, if a longitudinal study in Sweden is generalisable, career choices have largely been formed by age 13, becoming progressively more difficult to change with age (Lindahl 2007 cited in UNESCO 2017).

All teachers play a part in influencing STEM choices. Female teachers in STEM subjects provide role models for girls which helps improve girls' and women's self-perceptions and attitudes about STEM, so that STEM careers become an acceptable choice (Su et al. 2009; UNESCO 2017). Teachers will tend to reflect their own subconscious stereotypical views about who should or should not study STEM which can be passed on in class and influence their students' choices. A survey in the UK found that a significant percentage of male teachers (29%) deemed STEM careers more appropriate for boys than girls (Engineering UK 2018). Stereotyping extends beyond gender into ethnicity. Research in the US shows that there are also negative stereotypes (held by both staff and students) about ethnicity and STEM, which negatively influenced girls of African-American origin towards mathematics (Campbell 2012; Pringle et al. 2012 cited in UNESCO 2017).

Teachers and their teaching methods can positively influence girls towards STEM. Research shows that when mathematics is taught by working in collaborative environments, hence building on girls' earlier socialisation, using an applied perspective rather than theoretical perspective, girls appear to show more interest, get better grades and exhibit stronger aspirations than boys do (Geist and King 2008; Halpern 2004; Wang 2012).

When studying STEM subjects, women can find themselves in a minority and can have feelings of not belonging ('you're not one of the boys') which can lead to demotivation and lower grades, which can lead

[11] The study referred to here is conducted every three years across 80 countries as part of the OECD's Programme for International Student Assessment (PISA) tests for 15-year-old students in reading, mathematics and science. http://www.oecd.org/pisa/ (accessed 28 February 2019).

to thoughts about abandoning their studies (Cheryan et al. 2015; Murphy et al. 2007; Stout et al. 2011).

4.3.2.2 Entering the Sector

Professional attitudes influence women's inclusion in the sector. While many engineers and economists are not against gender equity, these professionals often do not see the relevance of gender to their work: 'what's it got to do with power lines?' (Christian Michelsen Institute 1999). A concern related to international energy companies is that some professionals consider that 'equality of the sexes is a matter of local culture and political concern' (Skutsch 1998: 947) which could imply that when working outside of their own country they would not take steps to support women's advancement in the sector. A six-country study representing different geographical regions of the world, found that what men said about their support for gender equality did not necessarily match their behaviour which could result in them actively resisting equality measures (Barker et al. 2011).

The energy sector has a highly masculine image, with a lingering perception by both women and men, that working in the energy sector requires physical strength which can act as a deterrent to women (Clancy et al. 2001). Occupational trades in the energy sector that required more skilled work, such as metal workers, insulations specialists, plumbers and pipe-fitters, electricians, heating and cooling experts are also mostly male-dominated (Sustainlabour 2013). For example, within the EU in 2016, women made up only 3% of the construction workforce (Eurostat 2016), apparently this situation has remained largely unchanged for over a century (Clarke and Sahin-Dikmen 2021). However, not only has the need for physical strength been reduced by the mechanisation and automation, but many of the jobs in the sector do not fall into this category such as project planning, finance, information technology, human resources, administrative support, business development and marketing skills, and legal knowledge. The notions gained in formative years that boys have an inherent capacity for STEM subjects which girls lack seems to continue into assessing women as less competent to hold technical jobs than their male counterparts, even when women have the same or better qualifications and work experience (Baruah 2017).

For women with appropriate qualifications for working in the energy sector, finding a vacancy is the first challenge. Personal networks are considered a key source of information about employment opportunities. However, men tend to have much better professional networks than women resulting in men receiving more timely information about jobs (UNESCO 2015). It is suggested that a factor in women's stronger representation in the renewable energy sector compared to the other energy sector branches is that the former networks tend to be more open. If men hold to the notion that women are not suited to technical functions, then they probably will not pass on the information about vacancies (IRENA 2019).

Networking can help in other ways to enable women to build their careers and gain a greater sense of satisfaction in their work. An example of how networks support women professionals in the energy sector comes from the Oil and Gas Sector in Pakistan where, in 1998, a Gender Equity in the Oil and Gas Sector—Pakistan Petroleum Women's Network was established. When the network was set up there were just under 700 women employees in a total workforce of around 34,000. Women reported that the network provided them with the space to discuss their problems and to work on solutions with other women, raising their self-confidence followed by feelings of improved job performance and satisfaction (Lele 1998).

Once an opening is identified, there is some evidence to suggest that women tend only to apply if they meet all the stated job requirements whereas men tend to be more speculative applying even if they only partially meet the requirements (Asia–Pacific Gateway Skills Table 2015). At the interview stage for technical positions, women face unconscious discrimination. Evidence shows that an employer will subconsciously opt for a candidate who personifies in terms of the characteristics of age, race and sex someone whom they are familiar with and/or who has previously carried out the function competently (Riach and Rich 2006).[12] This subconscious bias is potentially problematic in the established oil and gas industries which are traditionally dominated by white men (McKee 2014). It is possible that a woman's reference might not be helpful in

[12] Anecdotal evidence supporting this statement comes from an Indian businessman in response to the argument that the reason for the lack of women in the energy sector is because the pool of appropriately trained women is small: 'When you have a woman CEO in the company, miraculously, that pool emerges, and you get more women. Women hire other women' (Alves 2018 cited in IRENA 2019: 44). This remark agrees with the research in Austria cited in the introduction to this chapter.

drawing attention to their appropriateness for a position. An analysis of 312 reference letters submitted in support of applications for a medical position found that, for male applicants, their referees, who were usually men, tended to emphasise research skills, publications and career aspirations whereas for female applicants, the emphasis was on teaching skills, practical clinical skills and personal attributes (Trix and Psenka 2003).

Gender stereotyping about the nature of work that women are willing to undertake, for example, servicing equipment in remote locations, can also act as a barrier to employment. Such attitudes see women as a homogeneous group and decisions are made for them about the nature of the work they are offered (McKee 2014) which can result in women being restricted to posts considered more feminine, such as administrative and support services. What can help overcome these 'mind bugs' is a sudden expansion of an industry as appears to have happened in the US oil and gas sector where the number of women working in the sector increased from approximately 48,900 in 2004 to around 78,400 in 2012 (McKee 2014). When large-scale energy facilities, including wind and solar, are land-based, then accommodation becomes less of an issue compared to when operating off-shore.

There can also be practical issues that women see as barriers such as the availability of childcare facilities and washrooms (Wheldon et al. 2014). There can be a lack of clothing and equipment that match women's bodies (Fielden et al. 2000; Clarke et al. 2015).

4.3.2.3 Retention of Qualified Personnel

Once we have women in the sector, the next problem is keeping them there. The indications are that in companies, when measured over a 10-year period, the retention rate for women in STEM fields is half the rate of men (USAID 2018).

If salary levels are a motivating factor in job choice, for women learning that they earn less than men across all occupational categories within the energy sector, although the RE branch might be slightly better, they may choose to go elsewhere (IRENA 2019). Antoni et al. (2015) suggest that other factors, such as age and experience, could account for the gender pay gap. However, evidence related to apprenticeships, where these factors would not be so relevant, counter that argument. For apprentices (at least in the UK), women's levels of pay are lower pay than men (a study reporting in 2016 cited an hourly average of £4.82 compared with £5.85) (Young Women's Trust 2016). Women reported that they are less likely than men to receive training as part of their apprenticeship and they are more likely to be out of work at the end of their apprenticeship (Young Women's Trust 2016).

Women working in engineering trades report sexist attitudes and a range of forms of harassment which even if not intentional, create a feeling of exclusion ('you are not one of the boys') (see for example: McFarland 2015; Young Women's Trust 2016; Norberg and Johansson 2020).

Academia also has its problems with inclusion. Women in STEM subjects report feelings of isolation and lack of friendship, as well as feeling excluded from social gatherings (Massachusetts Institute of Technology 1999; Rosser 2005). A survey of over 4000 academics from more than 80 fields encompassing the natural and social sciences, the humanities and medicine based in nine research-intensive US universities found that women are more likely than men to report feelings of being imposters in their field. People with this so-called imposter syndrome doubt their abilities and feel a fraud (Muradoglu et al. 2022). These feelings are particularly strong in women from racial and ethnic groups but less so for men from these groups. There is a correlation between imposter syndrome and research fields that are perceived to require 'brilliance' (defined as innate intellect) to succeed which includes STEM subjects. Imposter syndrome in this context can be seen as a manifestation of the perceptions referred to above that boys have an inherent capacity for STEM subjects. The negative feelings are linked to the visual representation of the research world in which most scientists deemed successful are white men.

Globally women still have the main responsibility within families for care of children, sick and older family members. If women are in employment, these care responsibilities can require either a career break or a switch to part-time work which can be detrimental to promotion or career advancement. A direct impact is the restrictions on travel to participate in conferences and professional meetings which provide networking opportunities and building reputations as well as keeping in touch with the latest techniques or scientific breakthroughs. For academics, participation in conferences contributes to building their publications list which can play a role in grant applications an important element contributing to job promotion. Writing proposals and the process of publication is demanding on time. Women cite more frequently than men that their reasons for leaving STEM careers are family-related issues and time

(Frehill et al. 2008). The Department of Science and Technology in India runs a Women Scientists' Scheme (WOS) which offers financial support, training and mentorship return to encourage women to return to scientific research and development (Varthana 2024). There are also ingrained cultural attitudes which can work against women, for example, employers can question the commitment of women who take multiple career breaks (UNESCO 2017). If women indicate a preference for group work, this can be interpreted as a lack of self-confidence which in turn is seen as a personality weakness and even as a lack of intelligence (Etzkowitz et al. 1994).

4.4 Engendering the Energy Sector

The energy system is not being transformed in terms of the diversity of participating actors. While creating interesting and lucrative job opportunities for a group of highly educated women, does it matter if there is gender parity in the workforce profile?

Three areas need attention: encouraging diversity in the people who study STEM subjects, increasing the diversity in the sector at all levels and getting them to stay.

Changing negative attitudes towards girls studying STEM requires a focus on both parents, particularly mothers, and girls. Parents participating in school activities is found to be particularly effective at providing more information about STEM and promoting more informed discussion between parents and secondary school students about career directions (Van Voorhis et al. 2013). Targeting information about STEM for girls can help promote the energy sector as more female friendly and dispel myths that STEM does not have a social dimension. The grey literature has a range of examples to strengthen girls' interest in STEM and show that they have the ability to be STEM professionals. When teenage girls encountered female role models during summer activities related to STEM careers, they were found to show a greater interest in STEM subjects, for example, girls-only interactive workshops led by female tutors from a variety of occupational roles, including as scientists (Sadler et al. 2012). Holding these activities on campus gave female school students the opportunity to experience life at university which is considered to have positively influenced women enrolling for STEM degrees (O'Leary 2017). Also, the use of social media has been successful in creating a more realistic perception that university campuses are a place for women.

Mentoring has been found to improve girls' and women's self-confidence, self-esteem and motivation for participation in STEM studies and careers. Mentors act as a source of information, such as about scholarships, and play a role in introducing girls and young women to networks (which are recognised as an important resource in finding jobs). A mentor who shares a similar socioeconomic or ethnic background to her mentee can provide support about specific obstacles encountered in STEM careers (Bystydzienski et al. 2015).

Governments have a role to play in changing the attitude of girls towards STEM. In France, the Ministry of National Education and the Ministry of Higher Education and Research have appointed 'Science Ambassadors' who join classes 'to share their passion about science' while serving as role models of women in science. A survey in 2015, of 2000 students who had experienced a Science Ambassador in their class found that there was now more interested in scientific careers with 75% expressing that view compared to 46% before the interaction (UNESCO 2017: 69). In India, the government, has launched a number of initiatives, including workshops and mentoring, to encourage girls to consider a career in STEM (Varthana 2024).

Universities can look at their teaching techniques to encourage a range of methods, such as student-centred, inquiry-based and participatory strategies, that can help improve girls' self-confidence and efficacy (Baker 2013; Hughes et al. 2013; OECD 2014). Both the University of California, Berkeley, and the Massachusetts Institute of Technology have reported increases in female students studying engineering and computer science by using more group projects (IRENA 2019). STEM foundation year programmes could enable applicants with qualifications in non-traditional subjects to change career direction (Clancy and Feenstra 2019). The government of India tried a more formal approach by setting quotas for the number of girls enrolling for engineering at the country's 23 institutes of technology (Ibrar 2018 cited in IRENA 2019).

Teachers may also need to upgrade their skills to be more gender-aware about behavioural differences. In mixed sex groups, teachers need to ensure that girls are encouraged to speak out since girls tend to avoid confrontation with their peers whereas boys are not afraid to vocally defend their views. Also, boys are more likely to adopt leadership roles than girls (Baker 2000; Leman et al. 2016).

Having clear goals and setting targets are found to be effective for increasing diversity within a workforce. In the US, a review of 118 companies in 2015 found that those with gender targets made the most tangible progress towards gender balance and equity, compared to companies without targets (McKinsey et al. 2015 cited in IRENA 2019). France has set a mandatory gender quota of 40% women on company boards which is considered a contributory factor in the country having globally, as of 2015, the third-highest percentage of women board members. The percentage of women with seats on the boards of French companies had increased from 9% in 2009 to 37.6% in 2015 (Lee et al. 2015). A follow-up survey in 2021 involving 423 companies found, taking women as a homogeneous group, that there continues to be progress in gender balance in terms of percentages employed but less positive progress, particularly for women of colour and with disabilities, progressing up the ladder (McKinsey et al. 2021). The report also gives insights into the experiences of lesbian and bisexual women[13] in their workplaces including being more likely than women overall to experience microaggressions, such as being interrupted or spoken over, as well as having their judgement questioned. Such behaviour can be addressed in several ways. Having a clear set of gender-sensitive policies which address sexual harassment and discrimination lays the foundation of a system, transmitted through training of all staff, that enables employees to report violations safely and without fear of negative consequences (Ernst and Young 2016). Translating targets into detailed gender action plans for units or departments, using disaggregated data on diversity to monitor progress, has also contributed to changing work cultures (USAID 2018). Organisations can make public commitments to specific actions to address gender equality, for example, managers will be held accountable for reaching targets (Orser 2001).

Employment quotas or scholarships for a specified group can also have a negative side. A perception may be created that an individual hired under a quota scheme, or the recipient of a special scholarship, is in an

[13] The authors say that when referring to 'lesbian and bisexual women', this denotes women who self-identify as bisexual, pansexual, gay, lesbian or homosexual. The authors also acknowledge that the study samples include transgender and gender non-binary individuals, however, the sample sizes are not sufficient to generalise about their experiences.

organisation not because of their qualifications but because they represent a particular targeted category. Reducing negative attitudes about a diversified workforce can be addressed through training programmes for employees at all levels to explain about how being gender-aware and socially inclusive brings benefits for all staff members. For example, male employees of the Botswana Power Company became more receptive to the company's gender policy when they realised that it includes elements which also benefit men, such as paternity leave (Clancy et al. 2016).

Companies can be more aware of the visual images they present of the energy sector and the subliminal messages they send which are inadvertently contributing to the gender bias in recruitment. For example, advertisements with photographs or pictures showing men in hard hats communicates the message that this is not a job for women. On the other hand, less subtle are advertisements in China where there are very explicit statements expressing a preference for men and any that specifically referred to women were often sexualised, with aesthetic requirements and marital status being part of the selection criteria (Human Rights Watch 2018). In academia, being conscious that awards which only go to white men contributes to a perception that this is the only group with the capacity to succeed in scientific research.

In languages which have female and male forms of nouns using both forms in job advertisements can help reduce the perception that technical positions are only for men. For example, the energy regulatory authority in Costa Rica, when advertising for engineers, uses 'ingenerios/as' rather than only 'ingenerios' (USAID 2018).

Women's decisions about applications for a job or promotion can be influenced by the employer's working conditions particularly in relation to flexibility in working hours (including working from home) that enables them to meet their family needs and obligations (USAID 2018). Women may also opt for part-time work. The salary implications of part-time work are obvious but there can be other negative aspects compared to full-time employment of reduced or no benefits (such as holidays, sick leave with pay, health insurance, contributory pension schemes, social security benefits, etc.) (ILO, undated). Career development can also be affected for part-time workers through lack of training and profiling opportunities, for example, through participating in conferences, as well as commitment to the job being questioned.

To help meet diversity targets, organisations need to develop strategies which encourage more women and other underrepresented groups to apply for jobs by understanding differences in application strategies. For example, to address women's tendency to only apply for a job if in their assessment they meet all the criteria, MIT increased the number of female academic staff within STEM departments by cold calling suitable female candidates and encouraging them to apply (O'Leary 2017). To this approach could be added that the people contacted should represent a diversity of backgrounds.

To change the negative perceptions of women as employees and the use of behavioural traits as promotion criteria requires a change of work culture to one that is more gender-responsive, which enables a woman or a man who opts for balancing work and family responsibilities, particularly for young children and elderly relatives, to meet promotion criteria (Mason and Goulden 2002). To change work culture, some researchers (see, for example, Greed 2000) have argued requires a critical mass of people within an organisation—although there is no agreement on the size of this critical mass. The word 'people' rather than 'women' is used here deliberately. Firstly, 'women' are not a homogeneous entity. There are women who do not recognise the existing 'rules of the game' as a barrier for women in the world of work and politics.[14] So, it would be unwise to assume that women are more likely to represent women's interests than men are—indeed there is little conclusive evidence to show that they do (Childs and Krook 2009 cited in O'Neil and Domingo 2016). Secondly, men are also part of the work culture and they should play their part in bring change.

The worldview of the public, civil society and private sector organisations within the energy system, including the way they see system actors, influences the translation and implementation of energy policy. These organisations are also involved in the collection and interpretation of data from households, agriculture, businesses and communities to be translated into policy. Policy in turn is implemented at the micro-level by these organisations.

[14] An example of a woman who did not subscribe to a women's equality position is Margaret Thatcher, the British Prime Minister from 1979 to 1990. She believed that women should be promoted based on merit and not due to actions which could be defined as discriminatory towards one sex over another. She is also often quoted as describing feminism as poison (Purvis 2013).

4.5 What Are the Outcomes of Women Working in the Energy Sector?

There are a range of expectations about the likely outcomes of women working in the energy sector which fall broadly within the umbrella term: empowerment. As was pointed out in Chapter 2, empowerment is a multifaceted concept and can mean different things to different people. This section reviews the evidence as to the nature of that empowerment.

At the practical level women working in the energy sector gain a whole range of skills, from the basics of literacy to gaining technical knowledge and expertise and learning how to run a company across a range of scales. They also earn an income. So, does this economic empowerment from working in the energy sector or having energy access for enterprises bring a transformation in gender norms and greater gender equality? The empirical evidence is very limited (Haves 2012 cited in IISD et al. 2016). Initiatives to promote energy access do provide the opportunity for economic empowerment, however, this does not automatically lead to agency. The first challenge is whether social norms allow women to enter employment outside of the home (O'Dell et al. 2014). If women work their income may be considered the property of male members of the household who will decide how this income is used preventing woman from being able to accumulate assets (Standal and Winther 2016). Anecdotal evidence related to the Mali MFP, referred to in Sect. 4.2, reports that some men were resentful of women's opportunities to be involved with the Platform and the changes this brought to their self-confidence and independence, leading to physical abuse (Sovacool et al. 2013). However, the opposite can also happen. There is evidence that when women become entrepreneurs, contribute financially to the household, husbands are more inclined to help with domestic chores, particularly as was pointed out in Chapter 3, when electric equipment is available. Improvements in personal relations are reported. The approach by Solar Sister to hold group meetings for its entrepreneurs contributes to changing male behaviour. Men report that they have a better understanding of the women and girls in their families and that their communication with women in their families and communities has improved (Soria et al. 2016). There is also evidence from the Mali MFP programme that women began to acquire land which has been the traditional preserve of men (Sovacool et al. 2013). As was pointed out earlier, land ownership can be required as surety when applying for bank loans so this would build women's agency in enabling them to make decisions about their enterprises.

There are also negative aspects to women entering the formal workspace such as tensions and feelings of guilt that they are not meeting gender norms in relation to their household responsibilities. As a consequence, the coping strategies of 'superwoman' attempting to respond to all demands can result in exploitative practices such as hiring a female domestic helper who can be underpaid or undocumented or an unpaid family member.

Women based in rural areas who become energy entrepreneurs, particularly for those selling energy products rather than services based on modern energy, may see these activities as supplementary to income from agricultural activities (Pailman 2016). This business model has consequences for the client base. Firstly, if the service operates on a seasonal basis products that the women are selling may not always be available as needed. Also, if expanding their business requires finding new clients outside of the immediate locale and social network, travel is potentially going to be more difficult if attention to agricultural activities takes priority or cultural norms do not allow it.

After participating in self-help groups and as entrepreneurs within the energy chain, women report examples of increased psychological power when their perception is changed of what they can do—they go out on their own, call meetings, sign documents and negotiate with officials (ENERGIA 2019). Similarly, there are examples of increased social power. For example, in Rwanda, women involved in solar VLEs became recognised as leaders in their communities and were asked to mediate in conflicts between villagers (EPRU (UCT) and IPA 2019).

Whether women gain political power because of participating in energy projects varies with context and economic empowerment might not be sufficient to allow women a greater say in their communities, other supportive interventions may be needed. Male attitudes must change as well. Women taking on new roles can initiate that change. Bulelavata, a small, remote village in the Western Solomon Islands, built a community-owned micro-hydro scheme. The project approach involved women in a variety of roles which required them to learn new skills which built their confidence enabling them to play a more active and vocal role in the community's activities. The men in the community saw this as a positive development (Bryce and Soo 2004). In many communities village committees are structures usually dominated by men. To try to break that convention, an energy access project developer might require women's participation in committees and meetings. However, it is not a given that women will be able to use their voice effectively or speak at all, since in some cultures women are expected to remain silent in public. To break this norm requires a two-prong approach. First women require their confidence building to speak in public which can be done as part of a project, while in India, participation in women's self-help groups has been found to be an effective space whereas in rural Nepal separate women's and men's committees give women more influence (MSSRF and CRT Nepal 2016). The second area of attention is that men need sensitising to facilitate women's participation.

Before starting strategies to increase women's involvement in energy access projects which challenge social norms, an assessment of the social/cultural setting, particularly in respect of gender norms and values, is necessary in the design phase to minimise resistance. A political economy analysis of the energy sector using gender analysis was found to be helpful for identifying synergies and potential bottlenecks (Clancy et al. 2016).

What women do with their income varies. While it is not uncommon to read that they invest the income in their families, however, old traditions and practices live on. For example, it is reported that in the state of Odisha, in India, women used their earnings to buy jewellery which is the traditional 'savings bank', providing a reserve for emergencies such as illness (MSSRF and CRT Nepal 2019).

4.6 Conclusions

This chapter has reviewed the evidence related to women earning their living from the energy sector either directly in terms of formal employment in an energy company or as an entrepreneur selling energy services or using energy to run a business. The evidence, in terms of focus, is divided between the Global South, where the attention is primarily on women running businesses, and the Global North, where the attention is on women working for an energy company. Nevertheless, there is a universal commonality—there is still a way to go in terms of gender equality in the energy sector in meeting aspirations to work in or benefit from the services energy enables. There are no universal solutions to improve gender equity in the energy sector. This is not surprising since context matters in respect of gender norms. A survey by IRENA indicates that different regions of the world have different ideas about what will work (for example, some regions favour training and others favour gender targets) (IRENA 2019).

What enables women to work in the energy sector? Firstly, freeing up their time and labour invested in domestic activities. This can be done by acquiring equipment, such as improved cookstoves, community services (such as pumped water and grain mills) and men contributing both sharing responsibility for household tasks and prioritising purchasing labour-saving devices. There is certainly anecdotal evidence related to time-saving through access to alternative energy sources. For example, women participants in a Biomass Briquette Programme in Malawi reported that when they no longer had to travel long distances to collect fuel, they had more time to dedicate to income-generating activities (Mabona 2001). However, these women also reported that the free time they gained was also important for socialising with their friends.

Secondly, challenging and changing gender norms towards studying STEM subjects and the stereotyping of occupations. A lack of the formal vocabulary of science and engineering does not render a person incapable of having ideas about fulfilling needs and aspirations.

Processes of acquiring energy technologies can enable women to participate in the energy sector. Project approaches can be more gender-sensitive—moving beyond seeing women as passive consumers and ensuring that women are properly represented in decision-making where they were previously ignored. The village-level hydro project in the Western Solomon Islands referred to earlier in this chapter which involved women and men in its commissioning and operation, found that the logistics committee was female-dominated while men dominated the technical committee. The project organisers took the radical step of giving the committees equal status which helped to challenge the communities gender norms (Bryce and Soo 2004).

Cooperatives can provide a useful arena for women to learn skills and benefit from the space to articulate concerns and needs, as exemplified by Windfang, or in a collective space, such as the WBSECES. However, being part of a cooperative should be an active choice rather than a condition of access to resources. In an evaluation of women's experiences of micro-credit schemes run by NGOs in Tanzania, it appears that, in order to qualify for loans, in some schemes women were required 'to form groups in order to act as collateral, to attend meetings and make payments in very short time cycles' (Sigalla and Carney 2012: 551). For some women this was a stressful experience. On the other hand, Solar Sister appear to have created a model which provides support for its women entrepreneurs by holding village-level 'sisterhood meetings' creating safe spaces where the women can meet, sharing information about how to solve problems, improve skills, discuss new ideas and build trust (Soria et al. 2016). The latter has contributed to the development of lifelong friendships in which women feel they can find support when needed. The women expressed an interest in having a more formal network which is the approach used by WBSECES. The men involved with group meetings have a more formal format sharing only business ideas and not discussing personal issues.

Gender norms continue to create higher barriers for women's participation as energy entrepreneurs than for men's participation (e.g. requirement of property as collateral for loans). There are solutions to these barriers starting with governments, particularly since the origins of many of these barriers are not in the energy sector, although the energy sector organisations can play a part. For example, in Nepal, the issue of women's land tenure was addressed through a solar irrigation project. Women who owned or jointly owned land could obtain a 10% discount on a solar irrigation pump. As a result, 77% of applicants were women, of which 82% came from households that had transferred land into the woman's name (Mukherji et al. 2017). As an alternative to banks, micro-credit programmes have been a source of finance for entrepreneurs. However, they are also not beyond criticism. It should be kept in mind that income poverty reduction is the primary aim of many of the programmes, particularly funded by international agencies, and not women's empowerment or gender equality. An ethnographic study of women entrepreneurs (not necessarily engaged as energy entrepreneurs) in Dar es Salaam offers interesting insights into the experiences of micro-credit programmes. Annual interest rates can be as high as 30%, often to be repaid weekly, with husbands required to be guarantors and who, in some cases, have to give permission for their wives to participate (Sigalla and Carney 2012). While this study does have examples of women with increased agency to make decisions about their lives and their children's futures and improved, more equal relations with their husbands, there are other examples which give a different picture. The choices about the nature of enterprises divide along gendered lines with women's businesses based on their traditional household care activities, such as food preparation, which can struggle to generate sufficient income to repay the loans and meet bills. The women also found themselves juggling running their enterprise while complying with the behaviours, expectations and responsibilities towards their immediate and extended families who act as the guardians of the culturally constructed role of women as wives and mothers in Tanzania. Women entrepreneurs faced resistance not only from their husbands but also members of the extended family including female members.

However, economic empowerment of women is a complex process which requires a number of other changes related to gender equality to happen. There is a substantial literature advocating for and promoting women as energy entrepreneurs. While attention is given to helping overcome practical barriers, such as access to finance, less attention appears to be given to challenging gender norms which can prevent women becoming energy entrepreneurs. As Vossenberg rightly points out: entrepreneurship cannot be considered simply as an economic practice but as a practice that is embedded in a specific and gendered environment (Vossenberg 2013: 13). There are assumptions that to be an entrepreneur requires certainly characteristics such as leadership, risk-taking, embrace change, endurance, unemotional, do not require support and conformity (Buttner and Rosen 1988). These characteristics are generally ones associated with masculine qualities, therefore, even though both women and men can have these characteristics, we unconsciously have doubts as to whether women have the capacity to be entrepreneurs (Ahl 2006). Men appear not to associate 'feminine' characteristics with entrepreneurship (Gupta et al. 2009 cited in Marshall et al. 2017: 150). Indeed, we unconsciously see the entrepreneur as a man, for example, the literature is full of specific references to female entrepreneurs but not so often to male entrepreneurs (Clancy 2016)!

There are two consequences arising from the perceptions of entrepreneurship being a masculine endeavour. First, if the social norms that define the characteristics of what it means to be a woman do not match those of the characteristics of an entrepreneur, some women may be reluctant to challenge these norms to become entrepreneurs. On the other hand, there are women who can live comfortably with multiple identities (in this case, simultaneously as a 'good' woman/mother conforming to social norms and as an entrepreneur), but when a woman aspires to be an entrepreneur she is described as needing help to overcome structural barriers, which once removed she will now live up to the masculine norm of entrepreneurship. However, the barriers for women to become (energy) entrepreneurs are not only structural and the characteristics of entrepreneurship are not inherent but are also rooted in power relations (Ahl and Marlow 2012). This brings us to the second consequence of the perceptions of entrepreneurship being a masculine endeavour. If men are harbouring subconscious doubt about women's capacities to be entrepreneurs and they are responsible for making decisions about allocating resources to energy entrepreneurs, they may be reluctant to allocate those resources to women. Research in the UK provides some evidence to support the concern about gender bias when applying for finance, compared to men, women may be required to have more capital, be charged higher interest rates and have more loans refused (Hertz 2011).

Returning to Ahl and Marlow's point about entrepreneurship and power relations, it is not only being identified as a woman or man which influences an individual's capacity to be entrepreneurs but other aspects of identity, such as ethnicity, religion, sexual orientation and migration status, that intersect to create different experiences of oppression and privilege. This gives a more complex picture of what influences women's capacity to become an entrepreneur which taking an intersectional approach to gender analysis would give a better understanding of achieving women's empowerment across its multiple dimensions. It is much needed in energy entrepreneurship research. Also, without challenging patriarchal societies which create privileges and oppression, it is unwise to assume that women becoming entrepreneurs will automatically lead to greater gender equality, women's empowerment and increased agency.

Running an enterprise is not risk-free nor is training a guarantee of future employment. Women are promoted as energy entrepreneurs for 'hard to reach places' with the argument that they can draw on their social networks to find clients. However, these networks may not be sufficient to maintain a business. Several women solar engineers who participated in the Barefoot College programme found that they did not always receive the fees intended both to pay the balance of the cost of an installed solar panel and maintenance since households could not afford the payments (Mininni 2022). Not all the women trained in the Grameen Shakti programme remained working within the programme. More than a quarter of trainees, interviewed as part of an evaluation of the programme, had found income-generating activities elsewhere, some claimed to earn more money than the women who still worked for the programme (Hemson and Peek 2017). It is, of course, likely that the training was a contributing factor in the women securing alternative employment.

There is also a tension between enabling women to become entrepreneurs by NGOs reliant on external funding compared to an enterprise as part of a value chain where the lead company is aware of and responds to changes in the global market. The methods of solar panel construction have changed significantly over the last 20 years with the appearance in the market of ready assembled SHSs of better quality and lower prices than can be made by self-assembly at the village level a model on which a number of projects, including one of the most cited examples Grameen Shakti, were based on. It was women who were engaged in the self-assembly, whose jobs disappeared, while men had responsibility for installation and maintenance who still had employment opportunities (Hemson and Peek 2017). Being part of an integrated value chain might make entrepreneurs aware of impending changes and have time to adjust.

In terms of employment in the energy sector, there is still a lack of equality in the number of women and men employed in technical positions. Why this situation is conceived as a problem and how to solve it depends on who is asking the question. For a woman who wants to contribute to addressing climate change through their work, employment in the energy sector does not necessarily require STEM qualifications. Companies need staff with a wide range of qualifications, skills and knowledge. The women engineers of Windfang were delighted when women with accountancy skills wanted to be part of the cooperative. There is increasing recognition of the benefits of a diverse workforce so companies are looking at their staffing profiles and new opportunities can open up. Companies can take positive actions to change their staff profile by conducting an organisational assessment which can, among others, look at recruitment and retention practices.[15] However, a basic cause of the numerical inequality in employment is that men continue to outnumber women studying STEM subjects at school and in higher education. To help change this imbalance, companies can work with educational institutes to help encourage more women to choose STEM and helping to organise programmes for mothers to show that STEM are good for girls—while keeping in mind that approaches need to take into account that mothers are not a homogeneous group and reflect the cultural norms

[15] A number of tools exist to show how to conduct gender organisational change (see, for example, EIGE 2016).

of their communities. Companies can point out to Ministries of Education about the impact of not drawing on a pool of talent has for their businesses which has implications for the economy.

However, we need to keep in mind that the nature of work in energy sector needs a broad set of skills. It appears, at least in Europe, that the green energy transition will create employment mostly in the construction industry (European Commission 2019) which is a sector employing few women. Within the EU, in 2016, only 3% of those employed in construction were women (Eurostat 2016), although we cannot say anything about the diversity of employees in terms of their other social characteristics. This situation can be changed. Many of the buildings that will need to be upgraded to improve energy efficiency, such as, schools, social housing, hostels, care homes, hospitals, libraries and museums are publicly owned. This puts municipalities in a strong position to diversify the workforce since they can specify employment conditions for construction workers including family-friendly policies, such as flexible working hours and care-leave, and working conditions which benefit all employees. These conditions can apply both in-house and for external contractors. Municipalities can offer apprenticeships as part of the renovation work liaising with vocational training institutes, women's organisations, professional bodies and unions to ensure that the intake is diverse. There is work to be done!

References

Addati, L., Bonnet, F., Ernst, E., Merola, R. and Wan, P.M.J. (2016). *Women at Work Trends 2016*. Geneva: International Labour Office (ILO).

Alves, G. (2018). *Anand Mahindra Bats for More Women in Boardrooms; Says 10 pc Hike in Girls' Education Can Up GDP by 3 pc*. 7 November. https://economictimes.indiatimes.com/magazines/panache/anand-mahindra-bats-for-more-women-inboardrooms-says-10-pc-hike-in-girls-education-can-up-gdp-by3-pc/articleshow/66525064.cms

Ahl, H. (2006). Why Research on Women Entrepreneurs Needs New Directions. *Entrepreneurship Theory and Practice* 30 (5): 595–621.

Ahl, H. and Marlow, S. (2012). Exploring the Dynamics of Gender, Feminism and Entrepreneurship: Advancing Debate to Escape a Dead End? *Organization* 19 (5): 543–562.

Antoni, M., Janser, M. and Lehmer, F. (2015). The Hidden Winners of Renewable Energy Promotion: Insights into Sector-Specific WAGE Differentials. *Energy Policy* 86: 595–613.

Arregui, G., Candela, J., Estrada, B., Medialdea, B. and Pérez, S. (2010). *Study on Employment Associated to the Promotion of Renewable Energies in Spain*. ISTAS.

Arunachalam, R.S. (2008). Micro-Finance and Innovative Financing for Gender Equality: Approaches, Challenges and Strategies. In Johnson, T. (ed.), *Small Change or Real Change?: Commonwealth Perspectives on Financing Gender Equality* (pp. 195). London: Commonwealth Secretariat.

Asia-Pacific Gateway Skills Table. (2015). *Women in Transportation Careers: Understanding Participation in Canada*. Vancouver, Canada: Asia-Pacific Gateway Skills Table.

Baker, D. (2000). *Teaching for Gender Difference. Research Matters—To the Science Teacher*. National Association for Research in Science Teaching. https://www.narst.org/publications/research/gender.cfm (accessed 28 February 2019).

Baker, D. (2013). What Works: Using Curriculum and Pedagogy to Increase Girls' Interest and Participation in Science. *Theory into Practice* 52 (1): 14–20.

Barefoot College. (2015). *Volunteer Info Pack*. Tilonia: Barefoot College.

Barker, G., Contreras, J.M., Heilman, B., Singh, A.K., Verma, R.K. and Nascimento, M. (2011). *Evolving Men: Initial Results From the International Men and Gender Equality Survey (IMAGES)*. Washington, DC: International Center for Research on Women (ICRW) and Rio de Janeiro: Instituto Promundo.

Baruah, B. (2015). Creating Opportunities for Women in the Renewable Energy Sector: Findings from India. *Feminist Economics* 21 (2): 53–76.

Baruah, B. (2016). *Creating Opportunities for Women in the Renewable Energy Sector*. Fourth Annual Conference on Transforming Development Through Inclusive Green Growth, 6–7 September 2016. Jeju International Convention Center, Republic of Korea.

Baruah, B. (2017). Renewable Inequity? Women's Employment in Clean Energy in Industrialized, Emerging and Developing Economies. *Natural Resources Forum* 41 (1): 18–29.

Batliwala, S. and Reddy, A. (2003). Energy for Women and Women for Energy (Engendering Energy and Empowering Women). *Energy for Sustainable Development* 7 (3): 33–43.

Berner, E., Gomez, G. and Knorringa, P. (2012). Helping a Large Number of People Become a Little Less Poor: The Logic of Survival Entrepreneurs. *European Journal of Development* 24: 382–396.

Bose, M. L., Ahmad, A. and Hossain, M. (2009). The Role of Gender in Economic Activities With Special Reference To Women's Participation and Empowerment in Rural Bangladesh. *Gender, Technology and Development* 13: 69–102.

Brush, C.G. and Elam, A. B. (2023). Clearing the Hurdles: Revisiting the Under-Performance Hypothesis for Women-Led VC Funded Firms. *Journal of Small Business Management* 1–36.

Bryce, D. and Soo, C.C. (2004). Bulelavata Women Speak. *ENERGIA News* 6 (2): 19–20.

Buttner, E.H. and Rosen, B. (1988). Bank Loan OfFICER'S PErceptions of the Characteristics of Men, Women and Successful Entrepreneurs. *Journal of Business Venturing* 3 (3): 249–258.

Bystydzienski, J.M., Eisenhart, M. and Bruning, M. (2015). High School Is Not Too Late: Developing Girls' Interest and Engagement in Engineering Careers. *The Career Development Quarterly* 63 (1): 88–95.

Campbell, S.L. (2012). For Colored Girls? Factors That Influence Teacher Recommendations into Advanced Courses for Black Girls. *Review of Black Political Economy* 39: 389–402. https://doi.org/10.1007/s12114-012-9139-1

Campos, F., Goldstein, M., McGorman, L., Munoz Boudet, A.M. and Pimhidzai, O. (2015). *Breaking the Metal Ceiling: Female Entrepreneurs Who Succeed in Male-Dominated Sectors*. Policy Research Working Paper 7503. The World Bank Group Africa Region (accessed 22 April 2022).

Cecelski, E. (2005). *Energy, Development and Gender: Global Correlations and Causality*. Report Prepared for the UK Department for International Development (DfID) KaR Research Project R8346 "Gender as a Key Variable in Energy Interventions." https://www.energia.org/assets/2015/06/42-Energy-development-and-gender-global-correlations-and-causality.pdf (accessed 3 May 2022).

Charmes, J. (2012). The Informal Economy Worldwide: Trends and Characteristics. Margin. *The Journal of Applied Economic Research* 6 (2): 103–132. https://doi.org/10.1177/097380101200600202

Cheryan, S., Master, A. and Meltzoff, A.N. (2015). Cultural Stereotypes as Gatekeepers: Increasing Girls' Interest in Computer Science and Engineering by Diversifying Stereotypes. *Frontiers in Psychology* 6 (Article 49): 1–8.

Childs, S. and Krook, M. L. (2009). Analysing Women's Substantive Representation: From Critical Mass to Critical Actors. *Government and Opposition* 44 (2): 125–145.

Chopra, D. (2015). *Balancing Paid Work and Unpaid Care Work to Achieve Women's Economic Empowerment*. IDS Policy Briefing. Brighton: Institute of Development Studies (IDS).

Christian Michelsen Institute. (1999). *WID/Gender Units and the Experience of Gender Mainstreaming in Multilateral Organisations: Knights on White Horses?* Oslo: Christian Michelsen Institute.

Cirera, X. and Qasim, Q. (2014). *Supporting Growth-Oriented Women Entrepreneurs: A Review of the Evidence and Key Challenges*. Innovation,

Technology & Entrepreneurship Policy Note 5. Washington, DC: The World Bank.

Clancy, J. (2016). *In the Light of What We Know: Gender and Energy Transformations*. Inaugural lecture, University of Twente, 24 November 2016.

Clancy, J. and Feenstra, M. (2019). *Women, Gender Equality and the Energy Transition in the EU*. Commissioned Study for European Parliament's Committee on Women's Rights and Gender Equality. Brussels: Directorate General for Internal Policies: Policy Department C: Citizens' Rights and Constitutional Affairs (Women's Rights & Gender Equality).

Clancy, J.S., Alvarez, A., Maduka, O. and Lumampao, F. (2006). *Enabling Urban Poor Livelihoods Policy Making: Understanding the Role of Energy Services*. Synthesis Report Prepared for DFID KaR R8348. London: Department for International Development.

Clancy, J., Kustova, I., Elkerbout, M. and Michael, K. (2022). *The Gender Dimension and Impact of the Fit for 55 Package*. In Commissioned Study for European Parliament's Committee on Women's Rights and Gender Equality. Brussels: Directorate General for Internal Policies: Policy Department C: Citizens' Rights and Constitutional Affairs (Women's Rights & Gender Equality).

Clancy, J., Mohlakoana, N. and Diagne Gueye, Y. (with Muchiri, L. and Shakya, I.). (2016). *Mainstreaming Gender in Energy Sector Practice and Policy: Lessons From the ENERGIA International Network*. In Commissioned Study for ENERGIA Gender and Energy Research Programme (pp. 1–74). The Hague: ENERGIA.

Clancy, J.S., Stienstra, M., Gregory, J. and Cornland, D. (2001). *Gender and Energy–Women's Concerns in Energy: Background and State of the Art*. Working Paper 1. Study Carried Out for the European Commission Research Directorate.

Clarke, L., Michielsens, E., Snijders, S., Wall, C., Dainty, A., Bagilhole, B. and Barnard, S. (2015). *'No More Softly, Softly': Review of Women in the Construction Workforce*. Report Prepared for the 'Raising the Bar' for the Representation of Women in the Construction Workforce. https://westminsterresearch.westminster.ac.uk/item/9968q/-raising-the-bar-for-the-representation-of-women-in-the-construction-workforce (accessed 25 November 2022).

Clarke, L. and Sahin-Dikmen, M. (2021). Why Radical Transformation Is Necessary for Gender Equality and a Zero Carbon European Construction Sector. In Magnusdottir, G.L. and Kronsell, A. (ed.), *Gender, Intersectionality and Climate Institutions in Industrialized States* (pp. 164–180). London and New York: Routledge.

Corniaux, C. (2003). *La filière lait et produits laitiers dans la région de Saint-Louis*. Saint-Louis, Senegal: ISRA/CIRAD-PSI.

CPRC. (2008). *The Chronic Poverty Report 2008–09: Escaping Poverty Traps*. Manchester: The Chronic Poverty Research Centre (CPRC).

Davies, R.M., Olatunji, M.O. and Burubai, W. (2008). A Survey of Cassava Processing Machinery in Oyo State. *World Journal of Agricultural Sciences* 43 (3): 337–340.

de Haan, A. (2016). Enhancing the Productivity of Women-Owned Enterprises: The Evidence on What Works, and a Research Agenda. Www.idrc.ca/sites/default/files/sp/Documents%20EN/report_enhancing_the_productivity_of_women-owned_enterprises.pdf (accessed 22 April 2022).

Dejene, Y. (2007). *Promoting Women's Economic Empowerment in Africa*. African Development Bank (accessed 3 May 2022).

Delfs, C. (2000). Case-Study: Women Producing Electricity From Renewables: From 'Project' to 'Market': The 'Windfang' Project in Germany. *ENERGIA News* 3 (1): 9–11.

Diekman, A.B., Brown, E.R., Johnston, A.M. and Clark, E.K. (2010). Seeking Congruity Between Goals and Roles: A New Look at Why Women Opt Out of Science, Technology, Engineering, and Mathematics Careers. *Psychological Science* 21: 1051–1057.

Dowse, S., Horton, K., Lele, D. and Sherk, S. (1999). *Women in Canada's Oil and Gas Sector*. Report Prepared for the Canadian International Development Agency's Oil and Gas Sector Programme, Pakistan.

Du Reitz, A. and Henrekson, M. (2000). Testing the Female Underperformance Hypothesis. *Small Business Economics* 14 (1): 1–10.

Dutta, S. Kooijman, A. and Cecelski, E. (2017). *Energy Access and Gender: Getting the Right Balance*. Report Prepared for State of Electricity Access Report (SEAR). Washington, DC: ESMAP, World Bank.

EIGE. (2012). *Review of the Implementation in the EU of Area K of the Beijing Platform for Action: Women and the Environment Gender Equality and Climate Change*. Luxembourg: Publications Office of the European Union.

EIGE. (2016). *Institutional Transformation: Gender Mainstreaming Toolkit*. Luxembourg: Publications Office of the European Union.

ENERGIA. (2019). *Gender in the Transition to Sustainable Energy for All: From Evidence to Inclusive Policies*. Synthesis Report of the Evidence Generated by the ENERGIA Gender and Energy Research Programme. The Hague: ENERGIA. https://www.energia.org/new-gender-in-the-transition-to-energy-for-all-from-evidence-to-inclusive-policies/ (accessed 9 March 2022).

Engineering UK. (2018). *Gender Disparity in Engineering Report*. https://www.engineeringuk.com/media/1691/gender-disparity-in-engineering.pdf (accessed 31 March 2022).

EPRU (UCT) and IPA. (2019). *Female Microenterprise Creation and Business Models for Private Sector Distribution of Low-Cost Off-Grid LED Lighting:*

Multiple Randomized Experiments. Research Report RA5 for the ENERGIA Gender and Energy Research Programme. The Hague: ENERGIA/HIVOS.

Ernst & Young. (2016). *Women in Power and Utilities Index 2016*. http://www.ey.com/Publication/vwLUAssets/ey-talent-at-the-table-women-in-power-and-utilities-index-2016/%24FILE/ey-talent-at-the-table-women-in-power-and-utilities-index-2016.pdf

Espinal, R. and Grasmuck, S. (1997). Gender, Households and Informal Entrepreneurship in the Dominican Republic. *Journal of Comparative Family Studies* 28 (1): 103–128.

Etzkowitz H., Kemelgor C., Neuschatz M. and Uzzi, B. (1994). *Barriers to Women in Academic Science and Engineering*. In Pearson Jr, W. and Fetcher, A. (eds.), Who Will Do Science? Educating the Next Generation. Baltimore, MD: Johns Hopkins University Press.

European Commission. (2019). *The European Green Deal*. Brussels: European Commission. https://ec.europa.eu/info/publications/factsheets-european-green-deal_en

European Commission. (2022). *2022 Report on Gender Equality in the EU*. Luxembourg: Publications Office of the European Union.

Eurostat. (2016). *Share of Men and Women in 20 Most Common Occupations*. European Labour Force Survey.

Eurostat. (2022). *Employed HRST by Category, Sex, Age and NACE Rev. 2 Activity*. https://ec.europa.eu/eurostat/databrowser/view/HRST_ST_NSECSEX2__custom_2007239/bookmark/table?lang=en&bookmarkId=8aa75a0b-2116-498b-bc9a-5f3c8db8b6e9&page=time:2020 (accessed 4 September 2022).

EY. (2014). *Talent at the Table*. Report Prepared by Ernst and Young Global Ltd.

Fackelmann, S. and De Concini, A. (2019). *Why Are Women Entrepreneurs Missing Out on Funding? Reflections and Considerations*. Report Prepared for the European Commission. https://www.eib.org/attachments/thematic/why_are_women_entrepreneurs_missing_out_on_funding_summary_en.pdf (accessed 20 May 2024).

Feenstra, M., Creusen, A., Kottari. M. and van Tulder, F. (2022). *Klimaatverandering en Gender: Aanbevelingen, Beleidsanalyse, Lexicon*. Report (in Dutch) Prepared for The Netherlands Ministry of Education, Culture and Science (OCW). Amsterdam: Atria. https://75inq.com/services/library/klimaatverandering-en-gender/ (accessed 22 July 2022).

Fielden, S.L., Davidson, M.J., Gale, A.W. and Davey, C.L. (2000). Women in Construction: The Untapped Resource. *Construction Management and Economics* 18 (1): 113–121.

Frehill, L.M., Di Fabio, N., Hill, S., Trager, K. and Buono, J. (2008). Women in Engineering: A Review of the 2007 Literature. *SWE Magazine* 54 (3): 6–30.

Froehlicher, M., Knuckles Griek, L., Nematzadeh, A., Hall, L. and Stovall, N. (2021). *Gender Equality in the Workplace: Going Beyond Women on the Board*. https://www.spglobal.com/esg/csa/yearbook/articles/gender-equality-workplace-going-beyond-women-on-the-board (accessed 22 March 2021)

Garimella, A., Shardul, M., Chaudhury, S. and Palit, D. (2015). Institution as the Catalyst for Productive Use of Electricity in Livelihood Cluster: A Case for Energy Plus Approach from Andhra Pradesh, India. Paper Presented at 3rd International Conference on *Micro Perspectives of Decentralised Energy Supply*, April 23–25, 2015, BMS College of Engineering Bangalore, India.

Geist, E.A. and King, M. (2008). Different, Not Better: Gender Differences in Mathematics Learning and Achievement. *Journal of Instructional Psychology* 35: 43–52.

GEM (Global Entrepreneurship Monitor). (2012). *Global Report 2011: Global Entrepreneurship Monitor*. London: Global Entrepreneurship Research Association.

GEM (Global Entrepreneurship Monitor). (2018). *Global Report 2017/18: Global Entrepreneurship Monitor*. London: Global Entrepreneurship Research Association. https://www.gemconsortium.org/report/gem-2017-2018-global-report (accessed 6 May 2022).

GEM (Global Entrepreneurship Monitor). (2021). *Global Report 2020/21: Global Entrepreneurship Monitor*. London: Global Entrepreneurship Research Association. https://www.gemconsortium.org/report/gem-202021-womens-entrepreneurship-report-thriving-through-crisis (accessed 8 June 2022).

GEM (Global Entrepreneurship Monitor). (2023a). *Global Entrepreneurship Monitor 2022/2023 Global Report: Adapting to a "New Normal"*. London: GEM.

GEM (Global Entrepreneurship Monitor). (2023b). *Global Entrepreneurship Monitor 2023/2024 Global Report: 25 Years and Growing*. London: GEM (accessed 20 May 2024).

Glemarec, Y., Bayat-Renoux, F. and Waissbein, O. (2016). Removing Barriers to Women Entrepreneurs' Engagement in Decentralized Sustainable Energy Solutions for the Poor. *AIMS Energy* 4 (1): 136–172.

Govindan, M., Murali, R. and Dholakia, D. (2019). *Women in Energy: Breaking Stereotypes and Inspiring Change*. New Delhi: The Energy and Resources Institute (TERI).

Greed, C. (2000). Women in the Construction Professions: Achieving Critical Mass. *Gender, Work and Organisation* 7 (3): 181–196.

Gupta, V.K., Turban, D.B., Wasti, S.A. and Sikdar, A. (2009). The Role of Gender Stereotypes in Perceptions of Entrepreneurs and Intentions to Become an Entrepreneur. *Entrepreneurship Theory and Practice* 33 (2): 397–417.

Haider, K., Khanna, M., Kotei, M., Kushnir, K., Singh, S. and Sridhar, T. (2019). *Micro, Small and Medium Enterprises-Economic Indicators (MSME-EI)*. An Analysis Note December 2019. Washington, DC: International Finance Corporation (IFC).

Halpern, D.F. (2004). A Cognitive-Process Taxonomy for Sex Differences in Cognitive Abilities. *Current Directions in Psychological Science* 13: 135–139.

Harackiewicz, J.M., Rozek, C.S., Hulleman, C.S. and Hyde, J.S. (2012). Helping Parents to Motivate Adolescents in Mathematics and Science: An Experimental Test of a Utility-Value Intervention. *Psychological Science* 23 (8): 899–906.

Hart, K. (1973). Informal Income Opportunities and Urban Employment in Ghana. *Journal of Modern African Studies* 11 (01): 61–89.

Haves, E. (2012). *Does Energy Access Affect Women? Beyond Anecdotes: A Review of the Evidence*. London: Ashden. https://www.ashden.org/files/pdfs/reports/Ashden_Gender_Report.pdf

Hemson, D. and Peek, N. (2017). Training and Integrating Rural Women into Technology: A Study of Renewable Energy Technology in Bangladesh. *Gender, Technology and Development* 21 (1–2): 46–62.

Hertz, N. (2011). *Women and Banks: Are Female Customers Facing Discrimination?* London: Institute for Public Policy Research. https://www.ippr.org/files/images/media/files/publication/2011/11/women-banks_Nov2011_8186.pdf (accessed 27 May 2022).

Hughes, R.M., Nzekwe, B. and Molyneaux, K.J. (2013). The Single Sex Debate for Girls in Science: A Comparison Between Two Informal Science Programs on Middle School Students' STEM Identity Formation. *Research in Science Education* 43 (5): 1979–2007.

Human Rights Watch. (2018). China: Job Ads Discriminate Against Women: Government, Companies Seek "Men Only," Hype Having "Beautiful Girls." Human Rights Watch News, April. https://www.hrw.org/news/2018/04/23/china-job-ads-discriminate-against-women (accessed 5 April 2022).

Hunt, A. and Samman, E. (2016). *Women's Economic Empowerment: Navigating Enablers and Constraints*. Research Report Prepared for the Development Progress Programme. London: Overseas Development Institute (accessed 3 May 2022).

Ibrar, M. (2018). IIT-Delhi Tops MHRD Mandate, Enrolls 16% Girls in All Its Courses. Times of India, 5 July. https://timesofindia.indiatimes.com/city/delhi/iit-delhi-tops-mhrd-mandate-enrols16-girls-in-all-its-courses/articleshow/64862109.cms

IDS and GIZ. (2019). *Unlocking the Benefits of Productive Uses of Energy for Women in Ghana, Tanzania and Myanmar*. Research Report for the ENERGIA Gender and Energy Research Programme. The Hague: ENERGIA/HIVOS.

IISD (Global Subsidies Initiative), BIDS, IRADe and Spaces for Change. (2016). *Gender and Fossil Fuel Subsidy Reform: Current Status of Research*. Scoping Study Report RA4. ENERGIA Gender and Energy Research Programme. The Hague: ENERGIA/HIVOS.

ILO. (Undated). *What Are Part-Time and On-Call Work?* https://www.ilo.org/global/topics/non-standard-employment/WCMS_534825/lang--en/index.htm (accessed 7 April 2022).

InforWit Research Consultants. (2013). *Case Study on Biogas Entrepreneurship, Kenya*. Report Prepared for the Africa Biogas Partnership Programme (ABPP). The Hague: HIVOS.

IRENA. (2013). *Renewable Energy and Jobs*. Abu Dhabi: The International Renewable Energy Agency (IRENA).

IRENA. (2019). *Renewable Energy: A Gender Perspective*. Abu Dhabi: The International Renewable Energy Agency (IRENA).

Ishengoma, E.K. and Kappel, R. (2006). *Economic Growth and Poverty: Does Formalisation of Informal Enterprises Matter?* Report No. 20 Prepared as Part of the GIGA Research Program: Transformation in the Process of Globalization. Hamburg: German Institute of Global and Area Studies (GIGA).

Jennings, J.E. and McDougald, M.S. (2007). Work-Family Interface Experiences and Coping Strategies: Implications for Entrepreneurship Research and Practice. *The Academy of Management Review* 32 (3): 747–760.

Johns Hopkins University, Babson College and ICRW. (2019). *Women's Energy Entrepreneurship: A Guiding Framework and Systematic Literature Review*. Research Report RA7, ENERGIA Gender and Energy Research Programme. The Hague: ENERGIA/HIVOS.

Kabeer, N. (2012). *Women's Economic Empowerment and Inclusive Growth: Labour Markets and Enterprise Development*. Discussion Paper No. 29. Centre for Development Policy & Research, School of Oriental & African Studies, University of London.

Kakani, M. (2014). *The Women Barefoot Solar Cooker Engineers (WBSCE) Booklet*. Tilonia, India: Barefoot College.

Kelkar, G. and Nathan, D. (2005). *Gender, Livelihoods & Resources in South Asia*. Paper for the Fifth South Asia Regional Ministerial Conference Celebrating Beijing Plus Ten2 Islamabad, Pakistan, 3–5 May 2005.

Kelley, D., Baumer, B., Cole, M., Dean, M. and Heavlow, R. (2017). *Women's Entrepreneurship: Global Entrepreneurship Monitor 2016/2017*. Report for the Global Entrepreneurship Research Association. https://www.gemconsortium.org/file/open?fileId=49860 (accessed 6 May 2022).

Kelley, D.J., Brush, C.G., Greene, P.G. and Litovsky, Y. (2013). *2012 Women's Report*. The Global Entrepreneurship Monitor. Report for the Global Entrepreneurship Research Association.

Khamati Njenga, B. (2001). Upasi Stove Project. In Misana, S. and Karlsson, G. (eds.), *Generating Opportunities: Case Studies on Energy and Women*. New York: UNDP. ISBN 92-1-26124-4.

Khan, H. (2001). Battery-Operated Lamps Produced by Rural Women. In Misana, S. and Karlsson, G. (eds.), *Generating Opportunities: Case Studies on Energy and Women*. New York: UNDP. ISBN 92-1-26124-4.

Klapper, L.F. and Parker, S.C. (2011). Gender and the Business Environment for New Firm Creation. *World Bank Research Observer* 26: 237–257.

Leaper, C., Farkas, T. and Brown, C.S. (2012). Adolescent Girls' Experiences and Gender-Related Beliefs in Relation to Their Motivation in Math/Science and English. *Journal of Youth and Adolescence* 41: 268–282.

Lee, L.-E., Marshall, R., Rallis, D. and Moscardi, M. (2015). *Women on Boards: Global Trends in Gender Diversity on Corporate Boards*. MSCI ESG Research (Accessed 5 April 2022).

Lele, D. (1998). Gender Equity in International Petroleum Projects: Women in the Oil and Gas Sector. *ENERGIA News* 2 (3): 10–11.

Leman, P., Skipper, Y., Watling, D. and Rutland, A. (2016). Conceptual Change in Science Is Facilitated Through Peer Collaboration for Boys But Not Girls. *Child Development* 87 (1): 176–183.

Lindahl, B. (2007). *A Longitudinal Study of Students' Attitudes Towards Science and Choice of Career*. Paper Presented at the 80th Session of the International Conference of the National Association for Research in Science Teaching, New Orleans, LA. Cited in: Tytler, R. (2014). Attitudes, Identities and Aspirations Towards Science. In Lederman, N.G. and Abell, S.K. (eds.), *Handbook of Research on Science Education* (Vol. 2. p. 91). New York: Routledge.

Lumampao, F.G., Mataga, R.S. and Parado, B.C. (2006). Energising Poor Women Entrepreneurs. *ENERGIA News* 9 (2): 18–20.

Mabona, M. (2001). Ndirande Nkhuni Biomass Briquette Program. In Misana, S. and Karlsson, G. (eds.), *Generating Opportunities: Case Studies on Energy and Women*. New York: UNDP. ISBN 92-1-26124-4.

Maduka, J.O. (2006). Smoke Gets in Their Eyes: The Women Fish Smokers of Lagos. *ENERGIA News* 9 (1): 12–13.

Marlow, S. (2002). Women and Self-Employment: A Part of or Apart From Theoretical Construct? *Entrepreneurship and Innovation* 3 (2): 83–91.

Marshall, M., Ockwell, D. and Byrne, R. (2017). Sustainable Energy for All or Sustainable Energy for Men? Gender and the Construction of Identity Within Climate technology Entrepreneurship in Kenya. *Progress in Development Studies* 17 (2): 148–172.

Mason, M.A. and Goulden, M. (2002). Do Babies Matter? The Effect of Family Formation on the Lifelong Careers of Academic Men and Women. *Academe* 88 (6): 21–27.

Massachusetts Institute of Technology. (1999). A Study on the Status of Women Faculty in Science at MIT. *The MIT Faculty Newsletter* IX (4). March 1999.

McFarland, J. (2015). Are There Jobs for Women in Green Job Creation? *Women & Environments International* 94/95: 22–25.

McKee, L. (2014). Women in American Energy: De-feminizing Poverty in the Oil and Gas Industries. *Journal of International Women's Studies* 15 (1): 167–178.

McKinsey & Company and LeanIn.Org. (2015). *Women in the Workplace.* McKinsey & Company and LeanIn.Org.

McKinsey & Company and LeanIn.Org. (2021). *Women in the Workplace.* McKinsey & Company and LeanIn.Org. https://womenintheworkplace.com/ (accessed 5 April 2022).

Mininni, G.M. (2019). *Can Energy Empower Women? A Case Study of the Solar Mamas' Programme in Rural Rajasthan, India.* PhD thesis, University of Keele, UK.

Mininni, G.M. (2022). The Barefoot College 'Eco-village' Approach to Women's Entrepreneurship in Energy. *Environmental Innovation and Societal Transitions* 42: 112–123.

Mohlakoana, N., Knox, A., Ranzanici, A., Diouf, M., Bressers, H., de Groot, J., Pailman, W., and Sanfelice, V. (2018). *Productive Uses of Energy: The Informal Food Sector in South Africa, Rwanda and Senegal.* Research Report RA2, ENERGIA Gender and Energy Research Programme. The Hague: ENERGIA/HIVOS.

Morton, M., Klugman, J., Hanmer, L. and Singer, D. (2014). *Gender at Work: A Companion to the World Development Report on Jobs* (English). Washington, DC: World Bank Group.

MSSRF and CRT Nepal. (2016). *The Gender Factor in Political Economy of Energy Sector Dynamics.* Scoping Study Report RA3 Prepared for the ENERGIA Gender and Energy Research Programme 'Building the Evidence Base for Improving Energy Interventions' Effectiveness by Taking'. www.energia.org. The Hague: ENERGIA/HIVOS.

MSSRF and CRT Nepal. (2019). *The Gender Factor in Political Economy of Energy Sector Dynamics.* Final Report RA3. ENERGIA Gender and Energy Research Programme. The Hague: ENERGIA/HIVOS.

Mukherji, A., Chowdhury, D., Fishman, R., Lamichhane, N., Khadgi, V. and Bajracharya, S. (2017). *Sustainable Financial Solutions for the Adoption of Solar Powered Irrigation Pumps in Nepal's Terai.* Colombo, Sri Lanka: CGIAR Research Program on Water, Land and Ecosystems (WLE). https://wle.cgiar.org/sustainable-financial-solutions-adoption-solar-powered-irrigation-pumps-nepal%E2%80%99s-terai (accessed 13 July 2022).

Munyeme, G. (1999). Visiting Lecture to the International Training Workshop on Stand-Alone Rural Electrification, Technology and Development Group, University of Twente, The Netherlands. 17 May–19 June 1999.

Muradoglu, M., Horne, Z., Hammond, M.D., Leslie, S.-J. and Cimpian, A. (2022). Women—Particularly Underrepresented Minority Women—And Early-Career Academics Feel Like Impostors in Fields That Value Brilliance. *Journal of Educational Psychology* 114 (5): 1086–1100. https://doi.org/10.1037/edu0000669

Murphy, M.C., Steele, C.M. and Gross, J.J. (2007). Signalling Threat: How Situational Cues Affect Women in Math, Science, and Engineering Settings. *Psychological Science* 18: 879–885.

Norberg, C. and Johansson, M. (2020). "Women and "Ideal" Women": The Representation of Women in the Construction Industry. *Gender Issues* 38: 1–24.

Nordman, C.J. and Vaillant, J. (2014). *Inputs, Gender Roles or Sharing Norms? Assessing the Gender Performance Gap Among Informal Entrepreneurs in Madagascar*. IZA Discussion Paper 8046. Bonn: The Institute for the Study of Labor (IZA). https://www.iza.org/publications/dp/8046/inputs-gender-roles-or-sharing-norms-assessing-the-gender-performance-gap-among-informal-entrepreneurs-in-madagascar (accessed 3 May 2022).

O'Dell, K., Peters, S. and Wharton, K. (2014). *Women, Energy and Economic Empowerment*. New York: Deloitte University Press.

O'Leary, M.B. (2017). *Closing the Gender Gap in Mechanical Engineering*. https://news.mit.edu/2017/closing-the-gender-gap-in-mit-mechanical-engineering-0731 (accessed 3 May 2022).

O'Neil, T. and Domingo, P. (2016). *Women and Power: Overcoming Barriers to Leadership and Influence*. London: Overseas Development Institute (ODI).

OECD. (2014). *Fostering Equity in Higher Education: Compendium of Practical Case Studies*. Paris: Organisation for Economic Co-operation and Development. http://www.oecd.org/education/imhe/Promoting-female-participation-in-STEM.pdf

OECD. (2015). *The ABC of Gender Equality in Education: Aptitude, Behaviour, Confidence*. Paris: Organization for Economic Co-operation and Development (OECD) (accessed 24 August 2023).

Orser, B. (2001). *Chief Executive Commitment: The Key to Enhancing Women's Advancement*. Gender Diversity Tool Kit—Resource No. 2. Ottawa: Conference Board of Canada.

Pailman, W. (2016). *An Explorative Study of the Synergy Between Social Enterprises and Local Micro-Entrepreneurs in the Provision of Off-Grid Clean Energy Access*. Master's Thesis, University of Cape Town, South Africa.

Pearl-Martinez, R. (2014). *Women at the Forefront of the Clean Energy Future*. Washington, DC: IUCN-USAID. http://genderandenvironment.org/resource/women-at-the-forefront-of-the-clean-energy-future (accessed 30 March 2022).

Pringle R.M., Brkich K.M., Adams T.L., West-Olatunii, C. and Archer-Banks, D.A. (2012). Factors Influencing Elementary Teachers' Positioning of African American Girls as Science and Mathematics Learners. *School Science and Mathematics* 112 (4): 217–229.

Purvis, J. (2013). What Was Margaret Thatcher's Legacy for Women? *Women's History Review* 22 (6): 1014–1018.

Riach, P.A. and Rich, J. (2006). An Experimental Investigation of Sexual Discrimination in Hiring in the English Labor Market. *The B.E. Journal of Economic Analysis & Policy* 6 (2): 1–6.

Rick, K., Martén, I. and Von Lonski, U. (2017). *Untapped Reserves: Promoting Gender Balance in Oil and Gas*. World Petroleum Council and The Boston Consulting Group, 12 July. https://www.bcg.com/publications/2017/energy-environmentpeople-organization-untapped-reserves.aspx

Robb, A.M. and Watson, J. (2011). Gender Differences in Firm Performance: Evidence From New Ventures in the United States. *Journal of Business Venturing* 27 (5): 544–558.

Rosser, S. (2005). Women and Technology Through the Lens of Feminist Theories. *Frontiers* 26 (1): 1–23.

Sadler, P. M., Sonnert, G., Hazari, Z. and Tai, R. (2012). Stability and Volatility of STEM Career Interest in High School: A Gender Study. *Science Education* 96 (3): 411–427.

Sahgal, N., Evans, J., Salazar, A.M., Starr, K.J. and Corichi, M. (2021). *Attitudes About Caste*. In Religion in India: Tolerance and Segregation. Pew Research Centre. https://www.pewresearch.org/religion/2021/06/29/attitudes-about-caste/ (accessed 01 February 2024).

Sanjay, S., Chainani, R.M., Bakshi, R. and Naswa, T. (2023). *The Informal Sector in the Indian Economy*. Sattva Knowledge Institute, Bangalore, Delhi and Mumbai, India. https://www.sattva.co.in/wp-content/uploads/2023/12/14122023_Primer_Informal-Economy.pdf (accessed 7 June 2024).

Satyasai, K., Satyasai, B. and Smita, N. (2014). Self-Help Groups as Drivers of Entrepreneurship: Evidence From Karnataka and Odisha States. *Agricultural Economics Research Review* 27 (Conference Number): 123–133.

Shankar, A., Elam, A. and Glinski, A. (2019). *Women's Energy Entrepreneurship: A Guiding Framework and Systematic Literature Review*. Research Report RA7, ENERGIA Gender and Energy Research Programme. The Hague: ENERGIA/HIVOS.

Shankar, A.V., Onyura, M. and Alderman, J. (2015). Agency-Based Empowerment Training Enhances Sales Capacity of Female Energy Entrepreneurs in Kenya. *Journal of Health Communication* 20 (Suppl 1): 67–75.

Sigalla, R.J. and Carney, S. (2012). Poverty Reduction Through Entrepreneurship: Microcredit, Learning and Ambivalence Amongst Women in Urban Tanzania. *International Journal of Educational Development* 32 (4): 546–554.

Skutsch, M.M. (1998). The Gender Issue in Energy Project Planning Welfare, Empowerment or Efficiency? *Energy Policy* 26: 945–955.

Solar Foundation. (2018). *National Solar Jobs Census 2017. A Review of the U.S. Solar Workforce*. Washington, DC: Solar Foundation. February.

Soria, L., Farley, K. and Glinski, A. (2016). *With Solar Sister, Forward We Go: A Qualitative Assessment Exploring How Solar Sister Brings Light, Hope, and Opportunity to Women in Africa*. Washington, DC: International Center for Research on Women. https://www.icrw.org/wpcontent/uploads/2016/10/Solar-Sister-Qualitative-Assessment.pdf (accessed 25 July 2022).

Sovacool, B., Clarke, S., Johnson, K., Crafton, M., Eidsnes, J. and Zoppo, D. (2013). Renewable Energy the Energy-Enterprise-Gender Nexus: Lessons From the Multifunctional Platform (MFP) in Mali. *Renewable Energy* 50: 115–125.

Standal, K. and Winther, T. (2016). Empowerment Through Energy? Impact of Electricity on Care Work Practices and Gender Relations. *Forum for Development Studies* 43 (1): 27-45.

Stout, J.G., Dasgupta, N., Hunsinger, M. and McManus, M.A. (2011). STEMing the Tide: Using in-Group Experts to Inoculate Women's Self-Concept in Science, Technology, Engineering, and Mathematics (STEM). *Journal of Personality and Social Psychology* 100: 255–270.

Su, R., Rounds, J. and Armstrong, P.I. (2009). Men and Things, Women and People: A Meta-Analysis of Sex Differences in Interests. *Psychological Bulletin* 135: 859–884.

Sustainlabour. (2013). *Green Jobs and Related Policy Frameworks: An Overview of the European Union and South Africa*. A Report Prepared as Part of the Project "Social Dialogue for Green and Decent Jobs. South Africa - European Dialogue on Just Transition" Jointly Funded by EuropeAid and The Congress of South African Trade Unions (COSATU). SA/21.060200-01-08. https://edmsp1.ilo.org/KSP/en/Details/?dn=WCMSTEST4_038427 (accessed 28 March 2022).

Trix, F. and Psenka, C. (2003). Exploring the Color of Glass: Letters of Recommendation for Female and Male Medical Faculty. *Discourse & Society* 14 (2): 191–220.

UNESCO. (2015). *UNESCO Science Report: Towards 2030*. Paris: United Nations Educational, Scientific and Cultural Organization (UNESCO).

UNESCO. (2017). *Cracking the Code: Girls' and Women's Education in Science, Technology, Engineering and Mathematics (STEM)*. Paris: United Nations Educational, Scientific and Cultural Organization (UNESCO).

University of Oslo, TERI, Seacrester Consulting and Dunamai Energy. (2019). *Women's Empowerment and Electricity Access: How Do Grid and Off-Grid Systems Enhance or Restrict Gender Equality?* Research Report RA1, ENERGIA Gender and Energy Research Programme. The Hague: ENERGIA/HIVOS.

University of Twente, University of Cape Town, MARGE and ENDA ENERGIE. (2015). *Productive Uses of Energy: The Informal Food Sector in South Africa, Rwanda and Senegal*. Scoping Study Report, ENERGIA Gender and Energy Research Programme. The Hague: ENERGIA/HIVOS.

USAID. (2018). *Practical Guide to Women in Energy Regulation*. Washington, DC: USAID.

Van Voorhis, F.L., Maier, M.F., Epstein, J.L. and Lloyd, C.M. (2013). *The Impact of Family Involvement on the Education of Children Ages 3 to 8: A Focus on Literacy and Math Achievement Outcomes and Social-Emotional Skills*. MDRC. https://eric.ed.gov/?id=ED545474 (accessed 1 September 2023).

Varthana. (2024). *6 Collaborative Projects to Empower Women in Science Careers in India*. https://varthana.com/school/women-in-science-careers-in-india (accessed 7 June 2024.)

Vossenberg, S. (2013). *Women Entrepreneurship Promotion in Developing Countries: What Explains the Gender Gap in Entrepreneurship and How to Close It?* Working Paper No. 2013/08. Maastricht, The Netherlands: Maastricht School of Management.

Wang, M.-T. (2012). Educational and Career Interests in Math: A Longitudinal Examination of the Links Between Classroom Environment, Motivational Beliefs, and Interests. *Developmental Psychology* 48: 1643–1657.

Waters-Bayer, A. (1985). *Dairying by Settled Fulani Women in Central Nigeria and Some Implications for Dairy Development*. ODI Pastoral Development Network Paper 20c. London: Overseas Development Institute (ODI).

Weber, A. and Zulehner, C. (2010). Female Hires and the Success of Start-up Firms. *The American Economic Review* 100 (2): 358–361.

WECF. (2015). *Access to Affordable Low-Cost Solar Water Heating Solutions as a Basis for the First Gender Sensitive National Appropriate Mitigation Action in Georgia*. http://www.wecf.eu/english/energy-climate/

Wheldon, A., Sharma, C., Haves, E., Wheldon-Bayes, S. and Dobbs, E. (2014). *Review of How the Work of Ashden Award Winners Impacts the Lives of Women and Girls*. Report Prepared for UK Department for International Development, London.

World Bank. (2007). *A la Recherche de l'Emploi - Le Chemin vers la Prospérité (Looking for Work—the Road to Prosperity).* Country Economic Memorandum. Report No. 40344-SN Sénégal. Washington, DC: World Bank. https://openknowledge.worldbank.org/handle/10986/2184 (accessed 14 April 2022).

World Bank. (2011). *World Development Report 2012.* Washington, DC: The World Bank.

Yee, D.K. and Eccles, J. (1992). Parent Perceptions and Attributions for Children's Math Achievement. *Sex Roles* 19: 317–333.

Young Women's Trust. (2016). *Making Apprenticeships Work for Young Women.* https://www.youngwomenstrust.org/apprenticeshipcampaign/making_apprenticeships_work_for_young_women

Zolin, R., Stuetzer, M. and Watson, J. (2013). Challenging the Female Underperformance Hypothesis. *International Journal of Gender and Entrepreneurship* 5 (2): 116–129.

CHAPTER 5

Engendering Energy Policy

This chapter examines the extent to which, at least at the time of writing, energy policy has been engendered. The chapter starts with an analysis of what a gender-aware energy policy should look like. Section 5.2 provides a review of progress in obtaining such a policy which is followed by an analysis of why energy policy remains gender-blind. The final section of the chapter looks at experiences with approaches to changing the situation.

5.1 What Would We Expect From a Gender-Aware Energy Policy?

Policy is a statement of intent or commitment to act in a certain way by an organisation, which includes the government and, in the context of the energy sector, utilities, private sector companies and CSOs. Organisations can operate at different levels: local, national, regional and international. Not all government policy requires legislation but can be, as it is for many organisations, a written statement or a published circular with announcements increasingly appearing via social media. A policy plays several roles including providing the framework of strategies and plans for the distribution of, access to and control over resources used to address identified problems, which are prioritised to reflect the values and principles of the organisation. The existence of a policy framework is no guarantee of implementation. Policy is usually presented as objective, neutral and

J. Clancy, *Engendering the Energy System*,
https://doi.org/10.1007/978-3-031-78965-6_5

value-free which is a notion that feminists have challenged particularly in relation to assumed gender neutrality. As has been shown in earlier chapters in this book, the outcomes of energy policy are generally often not gender-neutral. So, in this section we look at what we could expect from an energy policy that is gender-aware.

First let us start with defining what we mean by a 'gender-aware'[1] energy policy. The overall aim of a gender-aware energy policy is to contribute to gender equality. Fundamental principles would include recognising the implications of the policy for both women and men which are likely to be different, identifying and implementing interventions that address gender gaps and overcome historical gender biases in policies (Prebble and Rojas 2018). Table 5.1 gives examples of the gender dimension of some key policy issues in the energy sector. The table shows the multidimensional character of energy policy reflecting political, environmental, economic and social issues. Each of these dimensions has a gender element. A task of government, as part of the political process, is to reconcile conflicting and convergent societal interests across these different dimensions (Doyle and McEachern 2001). In respect of energy policy, the societal interests include reconciling gender differences and inequalities in relation to the use, production, provision and distribution along the energy supply chain. This aspect is reflected in the SDGs which stress the synergies between gender equality and energy. As was pointed out in Chapter 2, there is a strong link with energy supply supporting economic development which is the notion underpinning much of energy policy. As Table 5.1 illustrates how the form of energy supply governments chose to promote has gender implications. The National Gas Policy of Tanzania (2013) recognises the benefits using natural gas for cooking has on women's drudgery and the burden of firewood collection as well as their health (Prebble and Rojas 2018). However, natural gas is a fuel of which the combustion products have an impact on the climate which reminds us that an energy policy's environmental considerations are at multiple levels from global to local and encompass air, water and land. These elements can be linked. So, a policy to promote the use of biogas could benefit the air quality inside kitchens as well as providing a fertiliser which has implications for soil fertility but to produce the biogas requires water. A societal aspect of the energy services chain would

[1] There are several terms used to describe a policy which has mainstreamed gender: gender-aware, gender-sensitive and gender-responsive.

include promoting women's involvement in the chain both in terms of employment and their capacity to influence the chain governance.

Feenstra (2002) considered that the formulation and implementation of a gender-aware energy policy requires gender-disaggregated data on energy uses and needs the collection of which should be carried out in a participatory manner. Since the specifics of a policy will depend on the context, we can formulate some generic statements about what we might expect from a gender-aware energy policy (Dutta 2003; Clancy and Feenstra 2006; ENERGIA 2007). A gender-aware energy policy would:

- recognise that women and men have different energy dynamics (roles in the household, decision-making areas, energy needs, responses to crises or coping mechanisms);
- recognise that women are a disadvantaged group relative to men, further differentiated in terms of their socioeconomic characteristics and access to and control over the factors of production;
- make available energy technologies and services that match the contextual gender-differentiated dynamics;
- employ appropriate policy instruments (such as taxation) to provide an enabling environment for policy implementation.

Once a policy is in place, it needs to be implemented in a gender-responsive way which includes creating opportunities for women's involvement and sensitising men to enable and support that involvement and a commitment to gender equality. However, it is not only the policy content which needs to be gender-aware but also policymakers who, in order to support gender equality, have to be able to understand the implications from a gender perspective of a particular programme, project or policy for both women and men. Those workers who implement policy, programmes and projects, as well as those who advise policymakers, need the capacity to take gender into account as part of their work.

As part of that capacity, energy planners need to have a better understanding of the gender differences in energy needs, in particular, that women's needs extend beyond cooking. In this respect, a broader analysis of energy needs is not helped by SDG7's focus on electricity and cooking. Table 5.2 shows, using the adapted Moser framework described in Sect. 2.2, how different energy forms can contribute to meeting women's needs and interests. However, the table does not show the

Table 5.1 The gender dimension of key policy issues in the energy sector (ENERGIA 2007)

Issues	*Gender dimension*
Investment decisions Which sectors receive which levels of support?	Policy determines which energy sectors receive attention and support. For example, large-scale expansion of the electrical grid without support for household connections may bypass poor women. On the other hand, support for village-level initiatives focused on renewable energy sources may provide women with both new energy services and employment
Access and availability of energy supply	Availability can affect women and men differently (for example, black outs when preparing meals can mean more work for women). Elements of energy access (cost, physical distribution, new technologies, etc.) can also offer more benefits to men than women Access is also affected by the choice of technology supported by policies, hence impacting on men and women differently (for example, use of non-standard valves and regulator for LPG cylinders)
Tariffs and pricing Are there favourable tariff structures for low-income groups? Who makes decisions in the household about access?	Lifeline tariffs and specific tariffs for different areas may influence access for different groups in the population, particularly since women are over-represented among the chronically poor. Gender differences in both access to and control over household income mean pricing decisions about tariffs and connection fees can affect women and men differently
Infrastructure construction Who benefits? Who is disadvantaged?	Large-scale energy projects raise gender equality issues relating to both social impact (how are women and men differently affected by large-scale dislocations?) and to employment possibilities (do women benefit from the increased economic activity?)
Community participation strategies Does the energy sector use participatory approaches to planning and project implementation?	Specific attention to addressing gender inequalities, based around cultural tradition and education, is required if women are to participate along with men

(continued)

Table 5.1 (continued)

Issues	*Gender dimension*
Human resources planning Does the energy sector recognize women's specific contribution in policy making? Are women given opportunities to move beyond their traditional jobs?	Personnel strategies (including hiring, promotion and access to training) should include measures to increase women's contribution to policymaking and to ensure women's access to opportunities in non-traditional fields
Environmental issues Who is exposed to what sort of hazards?	Given women's work in the home, they are more vulnerable to certain energy related environmental problems such as cooking smoke. Men tend to be more exposed to environmental hazards while working in energy-related sectors (work in nuclear plants, exposure to dangerous chemicals or live electrical wires)
Finance Who has access to financing (e.g. credit)?	Lack of collateral for women limits their access to financial resources thereby limiting their opportunity to move up the energy ladder, or opportunity to use energy resources for productive end use

outcomes, nor that men would also benefit from access to these energy forms.

However, Table 5.2 is indicative. It is constructed by well-meaning academics and practitioners who may not be best placed to determine whether these needs and issues are what women from another context, as the indented beneficiaries, themselves want from energy policy. The way women and men can influence creating a more gender-aware energy policy is discussed in Sect. 5.4.

A gender-aware policy needs to be backed up with sufficient financial resources to make it gender-responsive, in other words policy is translated from intent into action. Indeed, the allocation of financial resources for specific policy can be a key indicator of whether a government's commitment to addressing an issue (in this case gender inequality) is 'rhetoric or reality' (Kabeer 2003: 220). In their gender analysis of energy policy frameworks, Prebble and Rojas (2018) were able to find, as of 2017, only one country, Zimbabwe, which made a public commitment to providing financial support for gender mainstreaming into energy policy.

Table 5.2 Examples of how different energy forms can contribute to meeting women's needs and interests (Clancy et al. 2003)

Energy form	*Women's needs and issues*		
	Practical needs	*Productive needs*	*Strategic interests*
Electricity	• Pumping water supplies—reducing need to haul and carry • Mills for grinding reduces drudgery • Lighting improves working conditions at home	• Increase possibility of activities during evening hours • Provide refrigeration for food production and sale • Power for specialised enterprises such as hairdressing and internet cafes	• Make streets safer allowing participation in other activities (e.g. evening classes and women's group meetings) • Opening horizons through cell phones, radio, TV and internet
Improved biomass (supply and conversion technology)	• Improved health through better stoves • Less time and effort in gathering and carrying firewood	• More time for productive activities • Lower cost for process heat for income generating activities	• Control of natural forests in community forestry management frameworks
Mechanical	• Milling and grinding reduces drudgery • Transport of water and crops	• Increases variety of enterprises	• Transport allowing access to commercial and social/ political opportunities

5.2 Gender-Aware Energy Policy: Where Are We?

Energy policy is assumed to be gender-neutral benefitting women and men equally. Beneficiaries are identified using gender-neutral terms such as 'households', 'citizens' and 'communities' (Clancy et al. 2007; Osunmuyiwa, and Ahlborg 2019). However, the energy policy of most countries is gender-blind and does not provide the supporting framework that is required to meet gender needs in energy and transform the energy sector (Clancy 2009). Such a planning approach misses issues that are of relevance to women and inadvertently discriminates, usually against women compounded by ignoring their intersectional identities.

For example, the evaluation of a rural electrification project in Tamil Nadu showed that men benefitted more than women since the electricity was used to run irrigation pumps substituting for oxen-drawn water but not for household water needs (Rengasamy et al. 2001). The care of the oxen was traditionally a task for men, whereas household water provision is a task for women. The men gained more free time when the number of oxen decreased, which they used for involvement in politics and improving their agricultural methods. However, electricity did not substitute for any of the women's tasks. Another example of the way the promotion of electricity inadvertently discriminates against women relates to income generation. Many of women's traditional income-generating activities use process heat (such as, food preparation and processing, beer brewing and pottery) (Woroniuk and Schalkwyk 1998) for which electricity is not the cheapest option. Whereas a more gender-aware policy for small enterprises would consider the need for an energy form more compatible with process heat generation, for example, promoting effective distribution networks for Liquefied Petroleum Gas (LPG) (Cecelski and Matinga 2014).

Nevertheless, there are signs of energy policies beginning to be more gender-aware, although they divide between a perspective that sees women as passive beneficiaries and one that sees women playing an active role in the energy system. Engendering energy policy has initially been more actively pursued in the South than the North as can be seen in a content analysis of energy policy frameworks existing in 137 countries,[2] which found that, as of 2017, 61 (32%) contained references to gender (Prebble and Rojas 2018). Just over half of the references were found in frameworks from Sub-Saharan Africa (SSA) while only four out of 43 frameworks from countries belonging to the OECD had references to gender. In Kenya, the 2004 Energy Policy had very few references related to gender mainstreaming, however, by the time of the revised policy in 2011 gender issues have been incorporated throughout (Clancy et al. 2016). In Tanzania, the National Energy Policy 2015 promotes the balanced participation of men and women in the energy sector (IDS and GIZ 2019). Similarly, in Ghana, both the National Energy Policy (2010)

[2] These countries included 31 from the Asia and Pacific region, 25 from Latin America and the Caribbean (LAC), 28 from Sub-Saharan African (SSA), four from the Middle East and North Africa (MENA), 30 member states of the Organization for Economic Co-operation and Development (OECD) and 19 non-OECD Eurasia countries.

and the SEforALL Action Agenda of Ghana (2015) seek to promote the participation of women in the formulation and implementation of energy interventions as well as for building women's capacity to enable their participation in the energy sector (IDS and GIZ 2019). In India, the draft 2017 National Energy Policy recognises the impacts that access to clean cooking has on the health and drudgery of women and children (University of Oslo et al. 2019). In Nepal, energy policy has evolved from initially considering women only in the context of the household as passive beneficiaries (2006 Rural Energy Policy), then in the 2013 National Energy Strategy has specific goals related to time and drudgery, while in 2016 the Renewable Energy Subsidy Policy sees women having a more active role in promoting productive uses of renewable energy (University of Oslo et al. 2019). However, the 2016 Policy objectives do not explicitly associate women as having a role in the energy supply chain. There is a tendency by policymakers, for example, in Uganda and Senegal, to see a link between women and specifically renewable energies, rather than other forms of energy (Clancy et al. 2016). The 2013 National Energy Policy of Nigeria also frames women's needs as primarily related to their care economy tasks in the household although it does state that one of its primary objectives is to 'promote gender sensitivity' (IISD et al. 2019: 62).

Terminology in the formulation of the energy policy varies, for example, in Nepal, used 'women's needs' rather than 'gender issues' (University of Oslo et al. 2019), while in Ghana, the National Energy Policy (2010) refers to 'gender issues' (IDS and GIZ 2019). These findings are in keeping with a study in Eastern and Southern Africa nearly ten years earlier which found that often gender mainstreaming is formulated as either vague objectives that are difficult to measure or are narrow, practical and welfare-oriented, rather than directly promoting women's rights or empowerment (Karekezi and Wangeci 2005).

The North is slow in catching up and has chosen a different discourse with a more social inclusion orientation and an emphasis on the human capital agenda rather than the hardware focus of the Global South as promoted by SEforALL and SDG5. The European Parliament, in 2016, declared that a gender dimension must be included in EU energy policies and policy implementations (Clancy et al. 2017). While in February 2022 the Parliament passed a resolution recognising that women still face structural and cultural barriers to participation in all aspects of delivering the energy transition (Clancy et al. 2022). However, the EU Green Deal,[3] which sets out the plan to make the Union economically sustainable, makes no direction mention of gender or women and men although it speaks about a 'just transition and inclusive for all'. Nevertheless, gender aspects appear in one of the funding streams for the Just Transition: The European Social Fund Plus (ESF+). The ESF+ provides funding for actions related to employment, education and skills building to promote social inclusion within the EU. The gender dimension underlying the ESF+ comes from the principles of the European Pillar for Social Rights,[4] which includes a commitment to gender equality. Even so, the language used on the webpage describing the ESF+ is gender-neutral[5]. In contrast, the EU's Climate Pact[6] does contain gender and intersectional references, for example, the planned actions aim to remove barriers due to personal characteristics, such as gender, age and disabilities. The Pact also aims to give a central role of debates about the future of Europe to participants who are more socially diverse.

At the national level of the EU member states, countries are required to establish a 10-year integrated national energy and climate plan (NECP) to enable the EU to meet its energy and climate targets for 2030. However, as of April 2021, only seven[7] of the 27 member countries mention gender linked to human rights and/or SDG5.

Nevertheless, even when energy policy is formulated in a gender-neutral way, women can benefit. The Free Basic Electricity Policy of the South African Government—although originally intended for household use—has been used by women and men to set up businesses. This subsidy targeted at low-income households has enabled people to have access to an energy carrier that they would not previously have been able to afford. In a survey of 136 Street Food Sector (SFS) entrepreneurs 50 respondents (74% women; 26% men) reported that access to the free electricity

[3] https://ec.europa.eu/info/strategy/priorities-2019-2024/european-green-deal_en (accessed 1 April 2021).

[4] https://ec.europa.eu/info/strategy/priorities-2019-2024/economy-works-people/jobs-growth-and-investment/european-pillar-social-rights/european-pillar-social-rights-20-principles_en (accessed 1 April 2021).

[5] https://ec.europa.eu/esf/main.jsp?catId=62&langId=en (accessed 1 April 2021).

[6] https://europa.eu/climate-pact/index_en (accessed 1 April 2021).

[7] Austria, Belgium, Finland, Lithuania, Portugal, Spain and Sweden.

subsidy had encouraged them to start their SFS business (University of Oslo et al. 2019).

5.3 Why Is Energy Policy Gender-Blind?

Clancy and Feenstra (2006) identified two factors that influence the gender blindness of energy policy: women's social position and the attitude of energy organisations to gender issues. These two factors are of course linked. Women's control over their own lives is generally less than that of men; men tend to dominate decision-making within households, in communities and organisations. Policymakers tend to be men and energy organisations both in the public and private sector, as well as civil society (such as NGOs dealing with energy) tend to be male-dominated, particularly in the professional posts. Similarly, important energy customers for utilities, such as the large-scale industries and agriculture, are also dominated by men in senior positions. Even in NGOs the 'traditional' job distribution exits with men in technical posts and, if involved at all, women in administration (even when women have engineering qualifications) (University of Oslo et al. 2019). Male-dominated structures result in men talking to men about energy issues. Consequently, the forums where issues are identified, and the formulation of potential solutions tend to have an inadvertent male bias. An example of the male bias in policymaking is given in Box 5.1 which describes a participatory process at the macro-level in Botswana to review the national Energy Policy. It was primarily men with technical backgrounds who took part in the review. The approach to the engagement in the review process by the Ministry responsible for women's affairs (BWAD) illustrates two aspects of the problem: energy organisations being predominantly technical and not recognising the need for a social dimension and gender people seemingly reluctant to engage with the energy sector. As an example of the latter, BWAD had developed a multi-sectoral tool for creating a gender mainstreaming strategy involving a gender approach and a 'Step by Step Guide to Gender Mainstreaming' which, as of mid-2006, only 3 sectors of the economy (education, health and agriculture) had implemented them (Kealotswe 2006). A possible explanation is that these sectors can be considered to be those where gender issues are most visible.

Women are universally underrepresented in political decision-making bodies at the international, national and local levels. A global survey in 2015 found that 10% of energy ministries were headed by women

(Prebble et al. 2015 cited in Prebble and Rojas 2019: 2). Where women hold such positions, gender tends to play a higher profile role in energy policy formulation and implementation. For example, in South Africa and Uganda, women have held senior posts in energy ministries and gender issues have tended to be more visible (Feenstra 2002). Female politicians may avoid policy decisions related to gender equality if it risks being labelled as 'feminist' which does not fit their wider political agenda (Wiliarty 2011). However, overemphasis on equality of numbers should be treated with caution. Firstly, the evidence that women are more likely than men to represent women's interests is inconclusive (Childs and Krook 2009 cited in O'Neil and Domingo 2016). Secondly, the burden for achieving a gender-aware (energy) policy is not the responsibility of (a relatively small number of) women. Men also have a responsibility for delivering gender equality for which they have to become more gender-sensitive and gender-aware. Thirdly, by focusing on having more women visible in the public sphere of policy-making, the experiences of women in the private sphere remain unchallenged. Fourthly, by a reduction of representation to a binary set of numbers overlooks the heterogeneity of the categories labelled 'women' and 'men'.

Feminist political economists have argued that it is policymakers' failure to consider activities, such as water collection, to maintain the household, as 'work' and in so doing overlooks its contribution to that economy that results in the lack of attention to reducing women's labour and saving their time (Bedford and Rai 2010 cited in MSSRF and CRT Nepal 2019). Another domain where women's contribution is overlooked by energy sector policymakers is in the informal economy which provides a range of goods and services and makes a significant contribution to the national economy. A World Bank study in Senegal found that the informal economy, in 2007, contributed 12% of the GDP (World Bank 2007 cited in University of Twente et al. 2015: 16). However, the informal sector is, for a variety of reasons (see Chapter 4), usually disregarded by governments. This means that official data are not generally available. Nevertheless, the ILO found that in more than half of the 44 countries surveyed, women made up the largest share of employees in non-agricultural informal labour (ILO 2011). As has been described elsewhere in this book, women's work in households and enterprises would benefit from energy access.

It is in a government's own interests to ensure that energy policy addresses women's work, both paid and unpaid, first as a commitment to gender equality. The mutual reinforcement of the SDGs underlines the need for a more integrated approach to policies in order to achieve objectives. Secondly, if the government wants to promote economic growth which would require an expansion of the labour force, not addressing the issue of women's time and domestic work is short-sighted. Women in Japan entering the labour force after Second World War have been credited with Japan's economic development—not only in terms of output but also diversification in the economy. (Shioda 1994). The availability of appliances for household chores, combined with a good electricity supply, freed up women's time which is credited as a significant factor in enabling Japanese women to enter the workforce.

Box 5.1 Engendering Energy Policy in Botswana (Kealotswe 2006)
A review of the Botswana Energy Policy took place in 2002. A participatory approach, involving consultations with individual actors and reviews in mini workshops, was used to identify policy issues. At least 35 energy sector actors were consulted, 50% of which were representatives from government, 25% from parastatals and another 25% from private sector entities. Most of the participants in the mini workshops were male tending to represent the technical, supply-side of the sector with no-one from planning or social sciences. The Botswana Women's Affairs Department (BWAD) was not invited to participate. The output of the consultations contained no mention of gender issues, although there was reference to respiratory diseases from burning fuelwood and wood harvesting. The review process was completed with a workshop, organized to review the output of the consultative process. The participants were 34 men and five women (one from the private sector and four from government (including BWAD). Not surprisingly gender issues did not appear in the recommendations for the revised energy policy. The Energy Affairs Department (EAD) did attempt, several times, to seek assistance from the BWAD to integrate gender into the policy document without any response. The draft revised energy policy document was rejected in July 2005 by the Botswana Cabinet because of the absence of a gender objective.

Energy utilities also have policies in respect of employment policy and their conditions of providing a service: the former can influence a more gender-balanced staff across all grades and professions while the latter determines whether discriminatory policies continue, for example, requiring physical assets such as land or house, which are often not in a woman's name, as a security for a connection. While there have been improvements in gender balance in the workforce of energy utilities, women continue to be in the numerical minority overall as well as in senior or technical positions that form the route to leadership positions. A study of women in the global oil and gas sector found that, as of 2017, women constituted 22% of the workforce which is a much lower percentage than in manufacturing, finance, education, health and social work. A slightly better situation is found in the renewable energy sector, where women make up around 30% of the workforce (IRENA 2019). A survey, carried out in 2021, to assess the gender balance in employment in the renewable energy sector of 10 Sub-Saharan Africa countries[8] found that, of the 32 companies responding, 27% of the members of boards of directors, 30% of chief executive officers (CEOs), 16% of vice presidents and 26% of middle- and lower-level managers were women (IFC 2022).[9] Interestingly, the per cent of women on boards in SSA is better than the G-20 country average (20%). However, 8 of the 32 companies had no women on their boards.

In part, the gender imbalance in leadership positions in the energy utilities can be attributed to the small numbers of women graduating with appropriate qualifications suitable for a career in the energy sector, for example, degrees in the fields of science, technology, engineering and mathematics (STEM). On average, women make up between 20 and 30% of the STEM student cohort, which is significantly lower than for the average of women in tertiary education (50–60% of the student cohort) (Rick et al. 2017).

5.4 How to Engender Energy Policy

Policy is developed and implemented at both international and national levels, although this does not necessarily occur in isolation and there can be a dynamic between the two. The development of policy at the national level is strongly influenced by events on the world stage, such as the SDGs, as well as support from multilateral and bilateral development agencies. One would really like to think that policy was not formulated

[8] Cameroon, Ethiopia, Kenya, Mali, Nigeria, Rwanda, Senegal, South Africa, Togo and Zambia.

[9] These figures are in agreement with another study of 29 energy utilities in 24 SSA countries (Prebble and Rojas 2019).

based on responses to outside pressure but that it resulted from a process of broad consultation as part of a democratic process. Indeed, a gender just energy policy considers that procedural justice requires contributions from all stakeholder groups, not only people regarded as experts (such as energy engineers) but also 'ordinary citizens' who should be representative of the social diversity of their country or locality. However, there appears to be a lack of intersectionally disaggregated data to make an analysis of who participates in consultations. An example comes from the European Commission which asked for reactions to the proposed Fit for 55 package.[10] The Commission states it received over 45,000 responses, although there is no disaggregation to show the demographics of the respondents (Clancy et al 2022). If governments provide information about the groups which actually contribute, they are generally defined in broad categories such as social partners, civil society and the general public (Michael and Hultman 2023). An analysis of the National Recovery and Resilience Plans (NRRPs), which form a key policy instrument supporting the green transition for the member states of the EU, found that seven[11] explicitly mention the inclusion of stakeholders that have a specific focus on gender issues. For instance, the Cypriot Plan used the National Mechanism for Women's Rights and the Czech Plan was submitted to the Government Council for Gender Equality (Marco et al. 2024).

However, energy policy formulation and planning generally seems to take place at the national level, entirely in a top-down manner. Indeed, the governance landscape of the energy transition is becoming more complex, with increased involvement of state and non-state actors operating at different levels, taking responsibility for delivering the transition with a range of functions and competences, resulting in a reallocation of decision-making powers (Hooghe and Marks 2003; Jeffery and Peterson 2020). A concern about the competences of the people responsible for working with energy end-users comes from research in the Netherlands

[10] The Fit for 55 package is a set of proposals to enable the European Union's member states to achieve a reduction in the Union's net greenhouse gas emissions by at least 55% by 2030 EU 2030 climate and energy framework. https://commission.europa.eu/strategy-and-policy/priorities-2019-2024/european-green-deal/delivering-european-green-deal_en (accessed 20 June 2024).

[11] Cyprus, Czech Republic, Estonia, Ireland, Lithuania, Netherlands and Portugal.

and Austria which shows that many do not understand the complexities of the social dimensions of end-users resulting in those who need advice or assistance being overlooked (Clancy 2024). In part this can be attributed to their education. For example, a course in Austria to train energy consultants who would be giving advice on improving household energy efficiency included nothing about the social characteristics, such as gender, age and migration history, of the people the consultants would interact with (Hausner et al. 2023).

Promoting a more gender-aware energy policy requires strategies based on an understanding of the barriers to achieving that objective. Payne (2011) considers that these barriers can be categorised as pragmatic, conceptual or political which would require different strategies to address them. The pragmatic barriers reflect the level of resources required for successful formulation and implementation of gender equality policies. Conceptual barriers are related to the lack of a clear understanding of the concept of gender: for example, confusing gender with biological distinctions and overlooking its social construction. The conceptual barriers explain the continuance of pragmatic barriers while, at the same time, they also both underpin and reflect political barriers. The political barriers are rooted in the gender inequality of participation in political processes at all levels. For example, financing participation in meetings which would allow for raising issues of concern can be a pragmatic barrier for women. To participate in international meetings, women's organisations in the Global South may rely on donors for finance. Frenova (2021) found that, at least for meetings related to the United Nations Framework Convention on Climate Change (UNFCCC), donors were reluctant to fund participation in meetings, particularly those requiring international travel.

Feenstra's framework of enabling conditions for developing a gender-aware energy policy show components of the enabling environment and policy formulation processes which can help overcome Payne's barriers (see Table 5.3) (Feenstra 2002). Creation of a gender-aware energy policy is not dependent on all the elements in the framework in place. Instead, when the elements are combined, they can be considered an accelerant of the process (Clancy and Feenstra 2006). Nor should the elements be considered of equal value under all circumstances. Depending on context, a specific element may have a greater influence than others.

Table 5.3 Feenstra's framework of enabling conditions for developing a gender-aware energy policy

Enabling environment for engendering energy policy	
Dimension	*Description*
a. Participatory planning	An approach involving a range of actors (including civil society) is considered more likely to create a greater opportunity for women's voices to be heard than traditional approaches to policymaking
b. Gender Methodology	Involves having a gender strategy, collecting sex-disaggregated data and conducting gender analyses to develop a gender-aware energy policy
c. Legislation on Gender Equality	Form and scope: e.g. Is gender equality enshrined in the constitution?
d. Political commitment	Putting pledges into practice: e.g. the existence of a National Gender Policy
e. Institutional support	This support can come from within government, for example a Ministry for Women's Affairs or a Gender Ministry, or from civil society, for example, NGOs active in gender and energy
f. Financial commitment	Allocation of sufficient resources to implement gender-aware policies
Formulation process towards realising a gender-aware energy policy	
g. Gender-disaggregated data	Sex-disaggregated data are available
h. Gender-mainstreaming	Recognise that policy (particularly for energy) is gender-blind. Does any ministry carry out gender budgeting? Does women's/gender ministry provide support?
i. Participation	Women and men as energy end-users are involved in the policy formulation process
j. Recognition of gender energy needs	A clear statement exists recognising (i) women's role in energy provision and use and (ii) that women and men have different practical, productive and strategic energy needs
k. Integrated energy planning	Is there any attempt to integrate supply and demand sides? Any cross-ministry cooperation: e.g. energy and health; energy and education; energy and small industry?

Source Feenstra (2002)

An analysis of the extent of gender mainstreaming in the energy policy of six SSA and three Asian countries,[12] found that while some of the countries had gender statements in their energy policy, many of the elements of Feenstra's framework were missing (Clancy and Mohlakoana 2020). As was quoted above, Nepal, one of the countries in the study, has made strides in creating a more gender-aware energy policy. Institutional commitment to gender, which could come from within or be external to government, was found to be fragmented. NGOs focusing on renewable energy, which could be considered the natural home of gender-aware technologies and approaches, were using gender-neutral terms such as 'people' and 'communities' and formulated women's energy needs primarily as 'cooking'—although this could be influenced by the priorities set by international agencies' financing priorities. There seemed to have been, except for Senegal, a lack of engagement by Ministries for Women/Gender with the energy sector and NGOs with an energy focus seemed not to engage with them. Whether the SDGs, including the work of UN Women in respect of SDG7,[13] will change the level and nature of engagement remains to be seen. In Kenya, the Women's Bureau was tasked with collecting sex-disaggregated data. However, the Bureau was understaffed and under-resourced with limited authority over other ministries where gender desks were reported to be non-functional. This seems to fit a global pattern and one not exclusive to the energy sector (Alston 2014). A similar situation occurred in Botswana, where the BWAD's was not able to support gender mainstreaming in the energy sector because they were short of human resources (Kealotswe 2006). Kealotswe (2006) considered that the lack of progress with developing a gender-aware energy policy in Botswana was that men in the energy sector did not know how to mainstream gender. Likewise, in Kenya, the 2004 Sessional Paper on Energy states that deliberate steps will be taken to integrate women in policy formulation and management of the energy section, but without any indication of a strategy for how this is to be achieved (Sengendo 2008).

As was pointed out above, household and care work are not included in the calculation of national income of accounts, therefore it is invisible

[12] Kenya, Botswana, Senegal, Tanzania, Zambia, Lesotho, Nepal, Sri Lanka and Philippines.

[13] https://www.unwomen.org/en/news/in-focus/women-and-the-sdgs/sdg-7-affordable-clean-energy (accessed 10 May 2021).

to policymakers. Making this work visible and so creating awareness of its importance to the economy comes through the provision of data:

> No data, no visibility; no visibility, no interest. (Huyer and Westholm 2001)

The key role of data in policy formulation raises the question: how are the data generated? Based on experiences in other sectors, such as agriculture, forestry and health, it could be expected that local-level organisations operating with field-level extension workers and using participatory approaches are well placed to provide data about the communities to be served. Participatory approaches might be more inclined to reflect gender needs in energy. However, local-level representation and participatory approaches in energy planning by governments are not the norm (Ahmed Bakhiet 2008) and ministries of energy tend not to have local representation. This could be due to the energy sector being dominated by technologists and macro-economists who do not necessarily see the relevance of social, including gender equality, issues as part of their remit. Instead, energy planning is often implemented in 'a paternalistic, top-down manner in which needs are assessed by professionals' (Skutsch 2005: 37) which means that there is limited understanding of the communities to be served (Clancy 2000). Data gathering tends to use standard questionnaires administered to the household head who is most likely male, and has limited knowledge of women's energy needs, hence contributing to the gender blindness of energy policy. An example comes from recent research in Chhattisgarh, India, which found that the State Renewable Energy Development Agency (CREDA) used only male officers for community mobilisation in villages where mini-grids are to be installed and they interacted only with male villagers (University of Oslo et al. 2019). In terms of a more socially aware methodology of official data collection an example comes from the UK, where the English Housing Survey (EHS), an annual survey in England[14] of people's housing circumstances (including household income and the building's condition and energy efficiency), uses gender-neutral language and does not pose questions to a 'head of household' but to a household member

[14] The UK consists of four nations (England, Wales, Scotland and Northern Ireland) which uses slightly different data collection methods and metrics; hence it is not possible to give an aggregate picture for the UK of household energy use.

who has been designated as the Household Reference Person (HRP) or his/her partner (Clancy et al. 2017). The HRP is required to provide answers to questions about all the individuals in the household. The survey also collects data about a range of intersectionality dimensions within a household such as composition, ethnicity, age and employment status as well as whether they are tenants (private/social) or owner occupiers.

Raising awareness of issues to get them onto the political agenda requires having a political voice for women to undertake advocacy to allow them to define their own needs and propose solutions. India offers a model on how to open the opportunity for a broader cross-section of society to influence policy. A change in the Constitution of India in 1992 enabled the local village planning committees (panchayat) to initiate energy planning activities including the use of renewable energy sources to generate electricity.[15] There is a legal quota for women in which at least one-third of the panchayat membership should be women. While this creates the opportunity for village-level energy planning to have a more gender-aware approach, women's scope for influencing decision-making also needs the skills to participate in such activities as well as men's willingness to allow them the space to contribute. Women's participation in decision-making challenges the accepted norms and customs of patriarchy which shape entitlements and are taken for granted, a situation Bourdieu (1978) refers to as 'doxa'. In other words, men are being asked to possibly give up entitlements, for example, male farmers being first to benefit from village electrification by providing pumped irrigation water whereas women would benefit from pumped drinking water for their families. Such decisions may lead men to resist women's participation in village committees.

Participation is not a simple process. Who determines participation and who sets the agenda are key questions in shaping the outcomes of involvement. Communities are not homogeneous. They consist of a complex set of political and social relations delineated in terms of class, ethnicity and religion. So, it is possible that women who are on village committees may not reflect the heterogeneity of the community but represent the existing powerful groups in a community. Nor is there any guarantee that women on a committee will favour decisions that meet women's needs. Those

[15] https://www.india.gov.in/my-government/constitution-india/amendments/constitution-india-seventy-third-amendment-act-1992 (accessed 25 June 2024).

in power do not surrender their privileges easily to those who have little or no power (Pease 2021). Sometimes, privileged people do act against their own interests. An example comes from India, where the government organised a campaign to ask the economically better-off to forego the universally applied subsidy on LPG cylinders to help fund access by the less well-off to have access to a cleaner cooking fuel. It is reported that more than 200 million people benefitted from this initiative (MSSRF and CRT Nepal 2019).

Influencing energy policy decisions can come through participation in political processes, such as elections. There is evidence to suggest that in India, women's participation in Community Based Groups (CBOs) has resulted in women being seen as a specific group of interest to politicians seeking their votes. The distribution of LPG connections to women in low-income households in the state of Uttar Pradesh and the promise of free household electricity and appliances in Tamil Nadu are considered to have electorally favoured the politicians making these offers (MSSRF and CRT Nepal 2019). Women can also be active in improving policymakers' awareness of gender and energy issues and shaping energy needs from their own perspectives. Although women have complained that when political actions became 'professionalised', men can take over the strategic positions in an organisation and so weaken women's opportunities to influence political agendas (Roehr 2001).

Nevertheless, to enable more women to have a voice will require their capacity to be built not only in terms of knowledge about energy systems but also about how to operate in political processes. Any training to build capacity needs to be responsive to the lived experience of the women who will be trained. Enabling effective participation can depend on attention to 'simple' organisational matters such as the timing of meetings. It may be tempting to hold evening meetings. However, this overlooks women's household commitments which can extend into the evening hours. Women in the Netherlands reported they did not have time to participate in the planning committees of local energy initiatives (Garcia Vázquez 2016). They were also reluctant to participate since they considered that the issues discussed would be technical and beyond their knowledge. Allowing women to bring their children to meetings, as well as providing food, has been found to increase participation (Listo 2020). In some communities, even when there is street lighting, social norms can make it unacceptable for women to go out in the dark (Kelkar and Nathan 2007). For broader social inclusion in policy formulation, written material needs to consider women's literacy levels and the use of local languages (Listo 2020).

An underexplored issue in gender and energy is how meaningful is women's participation in processes to engender energy, in other words, participation in a policy formulation process should be instrumental in achieving objectives and not merely symbolic. Indeed, there is a lack of empirical evidence that women's participation has an impact on policy decisions, at least within gender mainstreaming in climate change processes (Chattopadhyay and Duflo 2004; Annesley and Gains 2010).

However, it is not only women's capacity that needs building but also other actors in the policy process. Approaches to building capacity to engender energy policy are described in Sect. 5.4.1.

The blame for lack of progress in creating gender-aware energy policies does not lie entirely with men with technical backgrounds. Gender specialists have been criticised for their lack of engagement in the energy sector, failing to understand its nature and relevance to women (Clancy 2011). However, the stress on the interconnectedness of the SDGs, in the context of this book SDGs 5 and 7, appears to be changing the engagement by gender experts (see, for example, Glemarec et al. 2016; Bell et al. 2020).

The lack of gender-disaggregated data means that the sorts of issues that tend to enter the public arena, and hence form the basis of policy responses, are related to men's responsibilities (for example, the need for irrigation pumps) whereas women's issues (for example, the need for drinking water) are overlooked. Indeed, policy design and the anticipated outcomes are often made on the basis of assumptions. For example, it is not unusual to read that women can increase their influence within the household and community by improving their economic status (see, for example, Gates Foundation undated). However, women's income generation activities tend to lie in the informal sector which do not lead to large cash surpluses. Even when women's contribution to household income is significant, it is not a given that women have control over major purchases (Clancy 2002). As was pointed out in Chapter 4, the informal sector's 'invisibility' to policymakers means that another route of contact between women and those involved in policymaking is constrained. Informal businesses may also not have trade associations which misses out on a potentially lobbying organisation. In short, women lack both a formal voice in the community and an informal voice through participation in the various institutions and organisations linked to the energy sector.

Another route for contributing to energy policy formulation comes through research. Women are already disadvantaged in making a scientific contribution due to the low percentage of women opting for STEM subjects (see Chapter 4). Data are hard to find on women's participation in energy research. An indication of the level of women's involvement in energy research comes from the European Union's Horizon 2020 research programme in which it is compulsory to collect sex-disaggregated data about staff involved in the project but not disaggregated further to indicate levels of social inclusion. Women are nearly half of the researchers (41.6%) but play less of a role as project coordinators (22.98%) and as members of advisory groups, expert groups and evaluation panels, and as individual experts (25.18%) (European Commission 2022). All proposal submitted to Horizon 2020 are required to have a Gender Equality Plan (GEP) in place. However, it is not clear if there is any assessment of a GEP as part of a project's evaluation for funding, since its existence is accepted based on self-declaration, nor its operation in practice. Proposals are required to include a gender dimension unless the proposers are able to justify its non-relevance.

It is fair to say that these problems are not exclusive to the energy sector, nor are they caused by the energy sector. What they represent is the manifestation of wider issues related to political ideology, culture and tradition.

5.4.1 Gender Analytical Tools for Policy Formulation

Two factors which act as a barrier to engendering energy policy are policymakers' uncertainties about how to integrate gender into policies and there is a general limited capacity for mainstreaming gender in energy policy (ENERGIA 2008), as is clearly illustrated by the experience with the energy policy in Botswana (See Box 5.1). To address these two barriers requires a set of tools that guide the process. Indeed, when interest in gender mainstreaming in the energy sector was growing, initially in the Global South, Skutsch (2005) identified a lack of appropriate gender-analysis tools to respond to the specific data requirements of energy planning. Skutsch considered that the then existing tools, such as the Harvard Method,[16] did not ask the right questions for the

[16] For an explanation of the Harvard Method (aka the Gender Roles Framework), see Overholt et al. (1985). See also Chapter 2.

energy sector particularly in relation to who decides about which type of energy source and technology for which activity. Once the right tools are available, a cadre of people with the knowledge and skills on how to mainstream energy are needed.

5.4.1.1 Gender Budgeting

One of the first approaches to address how to engender policy was gender budgeting[17] (GB) which consists of a range of activities to analyse whether the government budget advances gender equality, in other words the budget is gender-responsive (Budlender 2009). Such an approach meets several dimensions of Feentra's Enabling Environment. The GB approach differs from conventional government audits, which focus on administrative efficiency by making the link with a specific objective (in this case gender equality) (Sharp 2000), as well as analysing the impact on a specific group rather than as an aggregate of the whole population (UNIFEM 2002). A gender budget exercise can be used to: (i) raise awareness and generate understanding of gender issues and the impacts of budgets and policies on those issues; (ii) hold governments accountable for their budgetary policies and commitments in compliance with their international commitments (for example, Beijing Platform for Action; SDG5); and (iii) in line with those commitments, have government budgets and policies changed and refined to promote gender equality (Sharp 2000).A GB analysis can be carried out at any government administrative level, as well as for a specific sector or programme. Rwanda has, at least as of 2024, a Gender Monitor Office which is a government body which includes among its responsibilities is to ensure that the government budget is gender-sensitive.[18]

However, while the power of gender budgeting as a tool is recognised, it has been criticised as being very dependent on specialised skills to understand government budgets and so not participatory for ordinary citizens including those with little influence such as the poor and women. There has also been an over-reliance on external experts who lack the

[17] The literature contains at least three terms to refer to gender budgets: 'Gender Budget Initiatives' (for example, Balmori 2003), 'Gender Responsive Budgets' (for example, Budlender and Hewitt 2003) and 'Gender Budget Audits' (for example, Sharp 2000). Here the term 'gender budgets' is used as a generic term encompassing these different terms.

[18] https://www.gmo.gov.rw/about-gmo/overview (accessed 25 June 2024).

contextual understanding of a national expert which is important when dealing with an issue like gender which is rooted in local culture (Balmori 2003).

Data availability is another issue which makes GBs difficult to carry out. A comprehensive analysis would require data not only disaggregated by sex but, to ensure a budget allocation which is socially just, also requires data disaggregated across a range of socioeconomic indicators, such as age, ethnicity, marital status, level of education, employment and land ownership (ADB 2013). If data are available at all, the quality and quantity may be questionable, particularly at local levels of government which would probably be true for energy since this is not a sector generally devolved to this administrative level.

5.4.1.2 Gender Audits

To address some of the criticisms of gender budgets, particularly the lack of participation by ordinary citizens, another approach, known as gender audits[19] (GAs) emerged (Moser 2005a). The approach started, rather than as a tool to support gender mainstreaming for policy, as a mechanism for an organisation to carry out an internal evaluation of how gender issues were addressed in its programming portfolio and internal organisational processes (Moser 2005b). The approach was taken a step further by adding an external evaluation of the organisation's policies, programmes and activities. GAs can be used to achieve a range of objectives including to identify and analyse: (i) the factors that hinder efforts to mainstream gender in policy; (ii) the specific ways in which gender issues are, or are not, addressed; and (iii) critical gender gaps in existing energy policy formulation and implementation. The overall aim of a GA process is the development of recommendations on creating a more gender-aware policy.

An audit can be held at any government administrative level, as well as at the sectoral or programme levels acting as a learning tool for individuals, units and organisations to identify strengths and weaknesses in promoting gender equality and to identify solutions. The International

[19] To avoid the association of the word 'audit' with some form of financial evaluation, which can be politically sensitive when the organisation is a government entity involved in an external evaluation, the term 'gender assessment' can be used. See, for example, the Asian Development Bank: https://www.adb.org/themes/gender/country-gender-assessments (accessed 2 May 2017).

Labour Office (ILO[20]) uses gender audits, which it considers to be a social audit, as part of its monitoring and evaluation of its approaches to gender mainstreaming, (ILO undated). The ILO GA experience is that an analysis at the implementation stage brought changes in working practices not only at the field level but also within the organisation, such as changes to working hours, to better reflect responsibilities outside the workplace (ILO undated).

There is no standard procedure for conducting a GA, although methods used tend to be qualitative, such as desk studies, checklists, structured interviews, case studies and focus group discussions. Moser (2005a) uses three concepts to assess an organisation's progress with gender mainstreaming: (i) evaporation (when good policy intentions fail to be followed through in practice); (ii) invisibilisation (lack of documentation about what is occurring in practice); and (iii) resistance (when gender power relations are used to block gender mainstreaming efforts). A programme or project could be analysed in terms of its gender mainstreaming in its programme design (inputs), in its implementation (outputs) and in terms of its outcomes (increased gender equality and women's empowerment) (Moser 2005b).

Moser's approach tends to focus on implementation at the project level rather than on policy formulation to create gender-aware policies. So, to fill the gap of approaches to mainstreaming gender into energy policies, a gender audit methodology (GAM) was developed by ENERGIA, the international network on gender and sustainable energy (Barnett et al. 2010). ENERGIA's GAM is more than a set of tools but forms part of a strategy for engendering energy policy which is influenced by Feenstra's Framework (Clancy and Mohlakoana 2020). Rather than a report generated externally to the government system and delivered to the Ministry of Energy, ENERGIA'S methodology requires the involvement of the Ministry from the start of the audit process working with an audit team composed of national experts, supported by facilitators whose origins are in the region, who can articulate the gender dimension of energy in the local context. The aim here is counter a perception in the Global South that gender mainstreaming is a notion conceived in industrialised countries which is at odds with local values and norms and is imposed

[20] The ILO is a United Nations agency mandated to set labour standards, develop policies and devise programmes promoting decent work for all women and men.

as a conditional part of development funding. Arising from this perception comes resistance to incorporating gender as a planning concept. The underlying assumption in ENERGIA's strategy is that the initial involvement by the Ministry builds a sense of ownership of the final report and that this, in turn, increases the likelihood of policy implementation.

When gender mainstreaming in energy policy first emerged in the Global South, the cohort of gender and energy experts was small. To build capacity ENERGIA developed a training programme.[21] The trained experts then become part of a national capacity capable of responding to requests by governments and other energy sector organisations for support in mainstreaming gender. Team members come from a range of energy sector organisations representing public and private sector plus civil society organisations (such as utilities, agencies, companies and NGOs), as well as a range of government departments (for example, the Finance Ministry and the Statistical Office) and academia. Bringing together a range of departments can help breakdown silo thinking in which the impact of access to clean energy forms on a range of issues beyond economic development, such as health, are easily overlooked. The members reflect a range of technical backgrounds and expertise such as a gender specialist, energy planner, budget analyst, statistician, economist and policy analyst. Local experts, thereby reducing any feeling that gender is an externally imposed notion. ENERGIA's GAM approach of using a group of gender experts researchers consider a 'soft incentive' which offer an alternative to 'hard incentives' of binding and enforceable measures, such as sex-disaggregated targets and indicators (Holvoet and Inberg 2014). Some practitioners have argued that soft incentives are more likely to bring significant changes in social institutions than hard incentives which are likely to result in procedural compliance.

The audit process can take several months. The ENERGIA audit methodology consists of eight steps which are summarised in Table 5.4. The process involves a gender analysis of energy policy documents including identifying the existence of two key components financial

[21] The training programme consists of five training modules, which are available in English and French, grouped together under the title: The Gender Face of Energy. The Concepts in Gender and Energy; Gender Tools for Energy Projects; Engendering Energy Policy; Gender and Energy Advocacy; Engendering Energy Project Proposal Development: Capacity Building of Organisations; Project Proposal Development. Each module is accompanied by a trainer's guide. As of June 2021, only modules 1 and 2 are available on the ENERGIA website (www.energia.org) (accessed 2 June 2021).

commitments and the availability of sex-disaggregated data. An assessment is made of the institutional capacity of ministries to implement gender mainstreaming strategies together with a stakeholder analysis to identifying those currently not involved in energy policy formulation who could support gender mainstreaming activities. The final step is the validation workshop in which a report of the gender analysis of existing energy policies and a Gender Action Plan (GAP), which includes proposals for how to develop strategies for gender mainstreaming in the national context, are presented to the Ministry of Energy. An audit could be taken further once the GAP is being implemented to check on progress and evaluate the need for adjustments (EIGE 2019).

Between 2005 and 2011 ENERGIA conducted gender audits in 20 countries in Africa and Asia. The audit process has been refined based on experience. The audit in India included a simplified gender budget analysis of the Ministry of New and Renewable Energy in which participants categorised different programmes based on an assessment of their effectiveness in terms of the degree of relevance in addressing women's needs rather than a more rigorous financial analysis (IRADe 2009). Such a method can be seen as a starting point for discussion and raising awareness about gender issues which allows the involvement of a broader range of stakeholders than being confined to those with a narrow range of specialised skills.

Attribution of the existence of policy elements to a single action is methodologically challenging. However, an assessment of ENERGIA's gender audit process, found strong evidence to link some gender mainstreaming into energy policy to gender audits in Senegal and Kenya (Clancy and Mohlakoana 2020). The audit process itself was also considered to have influenced organisations' employment policy and working practices, as well as helping overcome male resistance to measures related to gender equality. Several ministries of energy which participated in the audits have established gender desks or focal points. The Kenya Power and Lighting Company used a gender audit to develop a company strategic plan for mainstreaming gender into polices, programmes and practices (Clancy 2011).

5.4.1.3 Gender-Just Energy Policy Framework

More recently in academic research, the concept of a gender-just energy transition to sustainable clean energy has emerged (Feenstra and Özerol

Table 5.4 ENERGIA's audit methodology (Clancy and Mohlakoana 2020)

Step	*Process*	*Output*
1	Background review of national gender and energy enabling framework	Gender situational analysis: identification of gender organisations and of government directives, mandates, policies and legislation on gender Inventory of energy production/supply by source, energy demand and consumption by sector, the access and affordability of energy services; energy sector organisations including employment profiles; energy policy formulation and implementation processes
2	Energy policy review	Gender awareness of energy policy based on quick-scan document analysis using keywords Energy statistics, data and indicators—extent of sex-disaggregation
3	Gender budget analysis	Extent of budget allocation to energy programmes which recognise the differentiated benefits for women and men
4	Gender organisational assessment	Capacity assessment of the Ministry of Energy to mainstream gender Capacity assessment of gender institutions to support gender mainstreaming in the energy sector
5	Key stakeholders analysis	Analysing perceptions about gender and the national energy policy by key stakeholders in the energy sector Identifying the types of support that key stakeholders could provide to support gender mainstreaming in energy policy
6	Pre-validation—SWOT analysis	Identifying strengths, weaknesses, opportunities and threats (SWOT) of gender mainstreaming in energy policy
7	Gender and energy action plan (GAP)	Agreeing on goals, outcomes, indicators, outputs, activities and responsibilities
8	Validation and dissemination	Ownership and endorsement of the GAP by the Ministry of Energy and key stakeholders

Step 1 = Feenstra's Dimensions a, b, c, d, e; Step 2 = g, j; Step 3 = f, h; Step 4 = e; Step 5 = a; Steps 6 and 7 = b; Step 8 = d

2021). The concept uses an energy justice framework to support policy-making which brings together three functions that are usually carried out separately: conceptualisation, analysis and decision-making (Sovacool and Dworkin 2015). Energy justice is presented as consisting of three tenets: distribution (access to energy services and allocation of benefits and harms of the energy supply system); recognition (communities differ across a range of social characteristics such as, age, cultural, ethnic, racial and gender differences); and procedural (all groups have a right to participate in decision-making, so how decisions are made is important for ensuring social inclusion) (McCauley et al. 2013). Energy justice takes into account the context, for example the social determinants of energy consumption which are gendered and vary across time and location (Fuller and McCauley 2016) as well as along the whole energy chain of supply and distribution (Bouzarovski and Simcock 2017). An energy justice analysis could also look at the whole energy system since an injustice experienced at the household level (for example, who decides about the household's energy choices) can be the result of decisions and mechanisms operating elsewhere in the energy system. Viewing the energy justice framework through a gender lens provides a nuanced understanding of how the energy transition may unevenly distribute benefits, costs and risks, thereby producing new inequalities or exacerbating existing ones.

Figure 5.1 shows the framework Feenstra used to analyse the 2019 National Energy and Climate Plans (NEPCs) for Bulgaria, France, the Netherlands, Spain and Sweden to identify the extent to which policies address gender inequalities in energy access (Feenstra 2021). The analysis found that gender-disaggregated data was scarce and the policies formulation generally used a social inclusion approach rather than a gender mainstreaming approach, which would be unlikely to address gender-specific needs and challenges. An advantage of the framework is that it enables the identification of policy options which would help develop a gender-just energy policy.

The analysis of the National Recovery and Resilience Plans (NRRPs) of the member states of the EU referred to at the start of Sect. 5.4 used a modified version of the Feenstra framework introducing a fourth dimension: a Just Energy Transition (Marco et al. 2024). The intention is to underline the social equality aspect of the energy transition. The researchers concluded that not all countries included a gender perspective in their provisions for the energy transition. Those countries which did include some provisions that could be considered gender-aware were

TENETS	ELEMENTS	CRITERIA
Rectogenital energy justice	Energy users	**R1. Intersectionality of energy users** R2. Diversity of energy users' needs
	Energy Poverty	R3. **Intersectionality of energy** poverty
Distributive energy justice	Energy **consumption**	D1. Gender equity in access to energy services D2. Gender equity in support for energy consumers
	Energy **production**	**D3. Gender equality in STEM education** D4. Gender equality in the energy labour market
	Energy governance	**D5. Cross sectional integration of gender-** energy nexus **D6. Multi-level integration of gender**-energy nexus
Procedural energy justice	Energy **participation**	**P1. Inclusive representation of actors**
	Energy rights	P2. Inclusion of gender-energy rights in **legislation** p3. Inclusion of gender-energy rights in policy documents

Fig. 5.1 Gender-just energy policy framework (Feenstra 2021)

only partial in the elements included in their policies. The researchers attributed the lack of gender-aware policies to two possible factors. Firstly, policymakers are not aware of the gendered dimensions of the energy transition and hence see no imperative to act. Secondly, which contributes to the lack of awareness is the lack of gendered disaggregated data.

Another version of Feenstra's framework was used to analyse the European Union's Fit for 55 package referred to at the start of Sect. 4.4 (Clancy et al. 2022). Again, the researchers used different elements and criteria within the three components which demonstrate the flexibility of the framework so that it can be adapted to the circumstances under analysis. This analysis was critical of the limited input women appear to be making into policy formulation.

5.5 Conclusions

In terms of progress in engendering energy policy, there are pluses and there are minuses. On the positive side, there are positive signs with more countries having some recognition of gender issues in energy policy. The South seems to have made more progress than the North in this arena, although interest in the North is growing. On the negative side, as the work by Prebble and Rojas (2018) shows, few countries have integrated gender into their policy frameworks. Indeed, the 27 member states of the European Union have paid little heed to the requirement for a gender dimension in their National Energy and Climate Plans (Clancy et al. 2022). There seems to be a time lag between mainstreaming actions and policy change which appears not to be a phenomenon confined to the energy sector. For example, Combaz (2013) reports on similar time lags in climate change processes. Another negative, is that where gender is integrated into energy policy the focus is very much in the Global South on addressing practical needs at the level of making tasks lighter and shorter or, in the case of the Global North, focusing on household ambient temperatures, important as these issues are, they are limited in focus. Energy policy does not readily directly address gender equality, even in the European Union, where all policies should comply with the European Pillar for Social Rights which includes gender equality as one of its principles. These examples illustrate the point made in Sect. 5.1 that the existence of a policy framework is no guarantee of implementation.

To explain why the time lag between a policy statement appearing in the public domain and being implemented exists and how to overcome it, Payne's framework of political, pragmatic and conceptual barriers, is helpful. The political barriers to a gender-responsive energy policy arise partly because of the way in which the issue arrives on the political agenda. As has been outlined in this chapter, women are constrained in access to the paths to influence policy. So, it is primarily men who formulate and implement (energy) policy generally with a weak understanding of gender concepts, particularly the relational aspects, while many remain to be convinced of the relevance of gender mainstreaming to achieve the energy transition. In other words, the conceptual barriers underpin the political barriers. A greater sensitivity and awareness of the gender issues could be generated if the data are available which if we reformulate Huyer and Westholm's mantra cited in Sect. 5.4: with data we have visibility which generates the interest leading to action. So, Payne would argue that it is a pragmatic barrier, that is a lack of resources, in this case the lack of data, which contributes to the political barriers.

However, there is also a requirement for building capacity in how to operate in a political arena. ENERGIA's audit methodology assumed that involving a ministry of energy in the process and the hand-over of a GAP in a validation workshop would result in the ministry accepting ownership of the document and proceed to implement the recommendations. However, policymaking is more complicated than handing over of a document, no matter how well formulated involving all the key stakeholders. Policymakers may interpret the audit exercise as an overt criticism of not implementing general government policy as one key informant reported in an evaluation of ENERGIA's audit methodology (Clancy and Mohlakoana 2020). The ministry might well see the audit report, not as a document providing a foundation for action but a potential basis for censure by opponents and so take steps to prevent the document entering the public domain. Clancy and Mohlakoana (2020) suggested that the energy ministry might be more ready to accept the audit outcomes if they had already been through an internal evaluation process in which an organisation evaluates how gender issues are addressed in their policies and their internal processes. The internal evaluation was not included in the initial GAM adopted by ENERGIA. Such an approach has been shown to work. For example, the Rural Energy Agency (REA), in Tanzania, participated in a gender audit which resulted in the establishment of a Gender Desk, REA staff trained on gender issues and the preparation of REA Gender Strategy and Action Plan including development of more gender-aware indicators in the Monitoring and Evaluation framework prioritising women's needs in electrification projects, such as wells, schools and dispensaries (ENERGIA 2019).

Bringing together stakeholders while profiling as democratic and inclusive can be problematic, if groups or individuals to participate in processes are not familiar with each other or that there is a lack of trust between the parties. A gender budget exercise in Zimbabwe appears to have been challenging to conduct due to the antagonistic relations that existed between civil society organisations and the government (Muchabaiwa 2010). Influencing policy can be through providing evidence by responding to calls for information, writing briefing papers or participating in meetings either as a participant or an observer both of which create opportunities for lobbying. In summary, influencing policy outcomes requires an understanding of the context as well as the way processes work.

References

ADB. (2013). *Delivering Results and Gender: Promoting Women's Empowerment Through Gender Responsive Budgeting in Malaysia*. Manila, Philippines: Asian Development Bank (ADB).

Ahmed Bakhiet, N.Y. (2008). *The Walk Out of the Rural Kitchen: Towards Planning Energy Services for Sustainable Rural Livelihoods in Sudan*. PhD Thesis, University of Twente, Enschede, The Netherlands.

Alston, M. (2014). Gender Mainstreaming and Climate Change. *Women's Studies International Forum* 47: 287–294.

Annesley, C. and Gains, F. (2010). The Core Executive: Gender, Power and Change. *Political Studies* 58 (5): 909–929.

Balmori, H.H. (2003). Gender and Budgets: A Practical Tool to Advance Towards Equity. *Bridge Issue 12: Gender and Budgets*. Brighton, UK: Institute of Development Studies.

Barnett, A. Groverman, V. and Wille, C. (2010). *ETC Energy Access Evaluation—Final Report*. Brighton, UK: The Policy Practice.

Bedford, K. and Rai, S. (2010). Feminists Theorize International Political Economy. *Signs* 36 (1): 1–18.

Bell, S.E., Daggett, C. and Labuski, C. (2020). Toward Feminist Energy Systems: Why Adding Women and Solar Panels Is Not Enough. *Energy Research and Social Science* 68: 101557.

Bourdieu, P. (1978). *Outline of a Theory of Practice*. Translated From the French Original by Nice, R. Cambridge, UK: Cambridge University Press, pp260.

Bouzarovski, S. and Simcock, N. (2017). Spatializing Energy Justice. *Energy Policy* 107: 640–648.

Budlender, D. (2009). *Ten-Country Overview Report. Integrating Gender-Responsive Budgeting into the Aid Effectiveness Agenda*. New York: UNIFEM. http://www.gender-budgets.org/index.php?option=com_joomdoc&view=documents&path=suggested-readings/integrating-gender-responsive-budgeting-into-the-aid-effectiveness-agenda-reports&Itemid=587 (accessed 21 March 2021).

Budlender, D. and Hewitt, G. (2003). *Engendering Budgets: A Practitioner's Guide to Understanding and Implementing Gender-Responsive Budgets*. London: The Commonwealth Secretariat.

Cecelski, E. and Matinga, M. (2014). *Cooking With Gas: Why Women in Developing Countries Want LPG and How They Can Get It*. Report Developed by ENERGIA for the World LP Gas Association.

Chattopadhyay, R. and Duflo, E. (2004). Women as Policy Makers Evidence From a Randomized Policy Experiment in India. *Econometrica* 72 (5): 1409–1443.

Childs, S. and Krook, M. L. (2009). Analysing Women's Substantive Representation: From Critical Mass to Critical Actors. *Government and Opposition* 44 (2): 125–145.

Clancy, J.S. (2000). *Policies, Projects and the Market: Empowering Women*. University of Twente TDG Working Paper 104.

Clancy, J.S. (2002). *Blowing the Smoke Out of the Kitchen: Gender Issues in Household Energy*. SPARKNET Briefing Paper 2. https://research.utwente.nl/en/publications/blowing-the-smoke-out-of-the-kitchen-core-issues-in-household-ene (accessed 24 February 2022).

Clancy, J.S. (2009). *Late Developers: Gender Mainstreaming in the Energy Sector*. UK DSA Annual Conference. Colerain. 2 to 4 September.

Clancy, J.S. (2011). *Swimming in the Mainstreaming: ENERGIA's Experiences with Engendering Energy Policy*. Review Paper Prepared for ENERGIA.

Clancy, J. (2024 forthcoming). *Making an Inclusive and Gender-Aware Energy Policy*. Paper Prepared for the Gender and Energy Research Programme of the User-Centred Energy Systems of the IEA Technology Collaboration Programme.

Clancy, J. and Feenstra, M. (2006). *How to Engender Energy Policy*. A Technical Briefing Paper Prepared for ENERGIA. The Hague: ENERGIA/HIVOS.

Clancy, J.S. and Mohlakoana, N. (2020). Gender Audits: An Approach to Engendering Energy Policy in Nepal, Kenya and Senegal. *Energy Research and Social Science* 62.

Clancy, J., Daskalova, V., Feenstra, M., Franceschelli, N. and Sanz, M. (2017). *Gender Perspective on Access to Energy in the EU*, Study for the FEMM Committee of the EU Parliament, PE 596.816, Brussels.

Clancy, J., Kustova, I., Elkerbout, M. and Michael, K. (2022). *The Gender Dimension and Impact of the Fit for 55 Package*. Commissioned Study for European Parliament's Committee on Women's Rights and Gender Equality. Brussels: Directorate General for Internal Policies: Policy Department C: Citizens' Rights and Constitutional Affairs (Women's Rights & Gender Equality).

Clancy, J.S., Malik, F.U., Shakya, I. and Kelkar, G. (2007). Appropriate Gender Tools for Unpacking the Gender-Energy-Poverty Nexus. *Gender and Development* 15 (2): 241–257.

Clancy, J., Mohlakoana, N. and Diagne Gueye, Y. (with Muchiri, L. and Shakya, I.). (2016). *Mainstreaming Gender in Energy Sector Practice and Policy: Lessons From the ENERGIA International Network* (pp. 1–74). Commissioned Study for ENERGIA Gender and Energy Research Programme. The Hague: ENERGIA.

Clancy, J.S., Skutsch, M.M. and Batchelor, S. (2003). *The Gender-Energy-Poverty Nexus: Can We Find the Energy to Address Gender Concerns in Development*? DFID Project CNTR998521.

Combaz, E. (2013). *Impact of Gender-Responsive Budgeting*. GSDRC Helpdesk Research Report. London: Department for International Development (DFID).

Doyle, T. and McEachern, D. (2001). *Environment and Politics (2nd edition)*. London and New York: Routledge.

Dutta, S. (2003). *Mainstreaming Gender in Energy Planning and Policies*. Background Paper for Expert Group Meeting UNESCAP Project on Capacity Building on Integration of Energy and Rural Development Planning. June–July 2003.

EIGE. (2019). *Gender Audit*. Luxembourg: Publications Office of the European Union.

ENERGIA. (2007). *Engendering Energy Policy*. Module 3, The Gender Face of Energy Training Programme. ENERGIA.

ENERGIA. (2008). *Turning Information into Empowerment: Strengthening Gender and Energy Networking in Africa*. Report for TIE-ENERGIA Project. Intelligent Energy Europe.

ENERGIA. (2019). *Gender in the Transition to Sustainable Energy For All: From Evidence to Inclusive Policies*. Synthesis Report of the Evidence Generated by the ENERGIA Gender and Energy Research Programme. The Hague: ENERGIA/HIVOS. https://www.energia.org/new-gender-in-the-transition-to-energy-for-all-from-evidence-to-inclusive-policies/ (accessed 9 March 2022).

European Commission. (2022). *Horizon 2020 Dashboard: Cross Cutting Indicators 1*. https://webgate.ec.europa.eu/dashboard/sense/app/5e5d5ced-03e4-4fd8-9251-0417d6ef4f33/sheet/35a06b85-15c7-4253-ab82-c889f6 30e4c9/state/analysis (accessed 29 November 2022).

Feenstra, M. (2002). *Towards a Gender-Aware Energy Policy: A Case Study from South Africa and Uganda*. Master's Thesis. University of Twente, The Netherlands. https://essay.utwente.nl/58197/1/scriptie_M_Feenstra.pdf (accessed 15 April 2021)

Feenstra, M. (2021). *Gender Just Energy Policy: Engendering the Energy Transition in Europe*. PhD Thesis. University of Twente, The Netherlands.

Feenstra, M. and Özerol, G. (2021). Energy Justice as a Search Light for Gender-Energy Nexus: Towards a Conceptual Framework. *Renewable and Sustainable Energy Reviews* 138: 110668.

Frenova, S. (2021). Orchestrating the Participation of Women Organisations in the UNFCCC Led Climate Finance Decision Making. *Climate* 9: 135.

Fuller, S. and McCauley, D. (2016). Framing Energy Justice: Perspectives From Activism and Advocacy. *ERSS* 11: 1–8.

Garcia Vázquez, C.S. (2016). *Local Energy Initiatives: The Female Perspective*. Master's Thesis, University of Twente, The Netherlands.

Gates Foundation. (undated). *Women's Economic Empowerment*. https://www.gatesfoundation.org/equal-is-greater/ (accessed 31 July 2023).

Glemarec, Y., Bayat-Renoux, F. and Waissbein, O. (2016). Removing Barriers to Women Entrepreneurs' Engagement in Decentralized Sustainable Energy Solutions for the Poor. *AIMS Energy* 4 (1): 136–172.

Hausner, B., Karner, S., Tomasi, H. and Badieijaryani, A. (2023). *Energy Consulting: A Tool for Inclusion?* Users TCP by IEA.

Holvoet, N. and Inberg, L. (2014). Multiple Pathways to Gender Sensitive Budget Support in the Education Sector. Analysing the Effectiveness of Sex Disaggregated Indicators in the PAF and Gender Working Groups in (Education) Budget Support to Sub-Saharan Africa Countries. 14th EADI General Conference, 23–26 June 2014, Bonn.

Hooghe, L. and Marks, G. (2003). Unravelling the Central State, But How? Types of Multi-Level Governance. *American Political Science Review* 97 (2): 233–243.

Huyer, S. and Westholm, G. (2001). *Toolkit on Gender Indicators in Engineering, Science and Technology*. Prepared for the UNESCO Gender Advisory Board. http://gstgateway.wigsat.org/TA/data/toolkit.html (accessed 28 October 2001).

IDS and GIZ. (2019). *Unlocking the Benefits of Productive Uses of Energy for Women in Ghana, Tanzania and Myanmar*. Research Report for the ENERGIA Gender and Energy Research Programme. The Hague: ENERGIA/HIVOS.

IFC. (2022). *Women's Participation in the Renewable Energy Workforce in Sub-Saharan Africa: Identifying Barriers and Opportunities for Women as Leaders and Employees*. Washington, DC: International Finance Corporation (IFC).

IISD (Global Subsidies Initiative), BIDS, IRADe and Spaces for Change. (2019). *Gender and Fossil Fuel Subsidy Reform: Findings From and Recommendations for Bangladesh, India and Nigeria*. Research Report, ENERGIA Gender and Energy Research Programme. The Hague: ENERGIA/HIVOS.

ILO. (undated). *ILO Participatory Gender Audit*. Geneva: Bureau for Gender Equality, International Labour Office of the International Labour Organisation (ILO). Geneva: International Labour Organisation (ILO).

ILO. (2011). *Statistical Update on Employment in the Informal Economy*. Geneva: International Labour Organisation (ILO) (accessed 9 August 2021).

IRADe. (2009). *Gender Analysis of Renewable Energy in India: Present Status, Issues, Approaches and New Initiatives*. Leusden, The Netherlands: ENERGIA.

IRENA. (2019). *Renewable Energy: A Gender Perspective*. Abu Dhabi: International Renewable Energy Association (IRENA).

Jeffery, C. and Peterson, J. (2020). 'Breakthrough' Political Science: Multi-level Governance Reconceptualising Europe's Modernised Polity. *The British Journal of Politics and International Relations* 22 (4): 753–766.

Kabeer, N. (2003). *Gender Mainstreaming in Poverty Eradication and the Millennium Development Goals*. London: The Commonwealth Secretariat.

Karekezi, S. and Wangeci, J. (2005). *Engendering Power Sector Policy in Eastern and Southern Africa*. Report Prepared for KaR Research Project "Gender as a

Key Variable in Energy Interventions" (R8346). Department for International Development, UK.

Kealotswe, M.M. (2006). *Gender Mainstreaming in Botswana Energy Policy: Model for a Gender-Sensitive Energy Policy*. Master's Thesis, University of Twente, The Netherlands.

Kelkar, G. and Nathan, D. (2007). *Testing of Gender Tools for Energy Projects in India*. Report for ENERGIA. Leusden, The Netherlands: ENERGIA.

Listo, R.E. (2020). *Energy for Empowerment: The Role of Energy in Women's Organising and Empowerment in Urban and Peri-urban South Africa*. PhD Thesis, University of Queensland, St Lucia, Queensland, Australia.

Marco, C., Cloe, M., Serena, T., Lucio, P. and Chiara, V. (2024). *A Gendered Assessment of EU Policies for Sustainable Energy System*. Report D1.2 for gEneSys Project. https://doi.org/10.5281/zenodo.11566948 (accessed 26 June 2024).

McCauley, D., Heffron, R., Stephan, H. and Jenkins, K. (2013). Advancing Energy Justice: The Triumvirate of Tenets and Systems Thinking. *International Energy Law Review* 32 (3): 107–110.

Michael, K. and Hultman, M. (2023). *Sweden's Integrated Energy and Climate Plan: An Analysis*. Users TCP by IEA.

Moser, C.O.N. (2005a). *An Introduction to Gender Audit Methodology: Its Design and Implementation in DFID Malawi*. London: Overseas Development Institute.

Moser, C. (2005b). Has Gender Mainstreaming Failed? *International Feminist Journal of Politics* 7 (4): 576–590.

MSSRF and CRT Nepal. (2019). *The Gender Factor in Political Economy of Energy Sector Dynamics*. Research Report RA3. The Hague: ENERGIA.

Muchabaiwa, B.L. (2010). *Gender-Sensitive and Child-Friendly Budgeting in Zimbabwe*. In McNeil, M. and Malena C. (eds.), Demanding Good Governance: Lessons From Social Accountability Initiatives in Africa (pp. 109–135). Washington, DC: World Bank Institute. http://www-wds.worldbank.org/external/default/WDSContentServer/WDSP/IB/2010/07/06/000334955_20100706051828/Rendered/PDF/555460PUB0Dema1EPI1978968101PUBLIC1.pdf (accessed 23 September 2017).

O'Neil, T. and Domingo, P. (2016). *Women and Power: Overcoming Barriers to Leadership and Influence*. London: Overseas Development Institute (ODI).

Osunmuyiwa, O. and Ahlborg, H. (2019). Inclusiveness by Design? Reviewing Sustainable Electricity Access and Entrepreneurship From a Gender Perspective. *ERSS* 53: 145–158.

Overholt, C., Anderson, M.B., Cloud, K. and Austin, J.E. (1985). *Gender Roles in Development Projects: A Case Book* (p. 331). Connecticut: Kumarian Press.

Payne, S. (2011). Beijing Fifteen Years on: The Persistence of Barriers to Gender Mainstreaming in Health Policy. *Social Politics* 18 (4): 515–542.

Pease, B. (2021). *Undoing Privilege Unearned Advantage and Systemic Injustice in an Unequal World*. London: Bloomsbury/Earthscan.

Prebble, M. and Rojas, A. (2018). *Energizing Equality: Sub-Saharan Africa's Integration of Gender Equality Principles In National Energy Policies and Frameworks*. Washington, DC: International Union for Conservation of Nature (IUCN) Global Gender Office.

Prebble, M. and Rojas, A. (2019). *Energizing Equality: Unlocking the Potential of Women's Participation in Sub-Saharan African Energy Utilities*. Washington, DC: International Union for Conservation of Nature (IUCN).

Prebble, M., Gilligan, M. and Clabots, B. (2015). *Women's Participation in Global Environmental Decision Making: An EGI Supplemental Report*. Washington, DC: International Union for Conservation of Nature (IUCN) Global Gender Office (GGO).

Rengasamy, S., Devavaram, J., Prasad, R., Erskine, A., Balamurugan, P. and High, C. (2001). *Thaan Vuzha Nilam Tharisu: The Land Without a Farmer Becomes Barren - Policies that Work for Sustainable Agriculture and Rural Livelihoods in Virudhunagar District, Tamilnadu*. London: IIED.

Rick, K., Martén, I. and Von Lonski, U. (2017). *Untapped Reserves: Promoting Gender Balance in Oil and Gas*. World Petroleum Council and The Boston Consulting Group. https://www.bcg.com/publications/2017/energy-enviro nmentpeople-organization-untapped-reserves.aspx (accessed 4 August 2021).

Roehr, U. (2001). *Gender and Energy in the North*. Background Paper for the International Conference "Gender Perspectives for Earth Summit 2002: Energy, Transport, Information for Decision Making", Berlin, 10–12 January 2001.

Sengendo, M. (2008). *A Summary of the Gender Audit of Energy Policies and Programmes in Botswana, Kenya and Senegal*. Leusden, The Netherlands: ENERGIA.

Sharp, R. (2000). *The Economics and Politics of Auditing Government Budgets for Their Gender Impacts*. Hawke Institute Working Paper Series No 3. Magill, South Australia: University of South Australia.

Shioda, S. (1994). Innovation and Change in the Rapid Economic Growth Period. In Nakamura, M. (ed.), *Technology Change and Female Labour in Japan*. Tokyo: United Nations University Press, 213 pp.

Skutsch, M.M. (2005). Gender Analysis for Energy Projects and Programmes. *ESD* IX (1): 37–52.

Sovacool, B. and Dworkin, M. (2015). Energy Justice: Conceptual Insights and Practical Applications. *Applied Energy* 142: 435–444.

UNIFEM. (2002). Introduction: Gender Budget Analysis and Accountability to Women. In Judd, K. (ed.), *Gender Budget Initiatives: Strategies, Concepts and Experiences*. High Level International Conference 'Strengthening Economic and Financial Governance Through Gender Responsive Budgeting' Brussels,

16–18 October 2001. New York: United Nations Development Fund for Women (UNIFEM).
University of Oslo, TERI, Seacrester Consulting and Dunamai Energy. (2019). *Women's Empowerment and Electricity Access: How Do Grid and Off-Grid Systems Enhance or Restrict Gender Equality?* Research Report, ENERGIA Gender and Energy Research Programme. The Hague: ENERGIA/HIVOS.
University of Twente, University of Cape Town, MARGE and ENDA ENERGIE. (2015). *Productive Uses of Energy: The Informal Food Sector in South Africa, Rwanda and Senegal.* Scoping Study Report, ENERGIA Gender and Energy Research Programme. The Hague: ENERGIA/HIVOS.
Wiliarty, S. (2011). Gender and Energy Policy Making Under the First Merkel Government. *German Politics* 20 (3): 449–463.
World Bank. (2007). *A la Recherche de l'Emploi - Le Chemin vers la Prospérité.* Washington, DC: World Bank. Rapport No. 40344SN.
Woroniuk, B. and Schalkwyk, J. (1998). *Energy Policy and Equality Between Women and Men.* SIDA Equality Prompt #9. Stockholm: Infrastructure Division, SIDA.

CHAPTER 6

Transforming the Energy System: Looking Back to Move Forward

6.1 Introduction

In this chapter, I look back at some of the lessons we have learned in gender and energy research and practice in terms of reaching the objective of generating a gender-just energy transition. I have some suggestions for areas that need further research.

Section 6.2 gives a very brief overview of the way that gender and energy has evolved as a research area. Section 6.3 warns of the dangers associated with making assumptions about the outcomes of energy access and technological interventions. Section 6.4 explains what, how and when we count are important for ensuring we generate a gender-just energy transition. Finally, in Sect. 6.5 there are some suggestions for further research.

6.2 The Evolution of Unpacking the Gender and Energy Nexus

The gender and energy nexus has evolved considerably in terms of the way the subject is studied. Interested started more than 40 years ago when a social dimension (women's time) was linked to an environmental issue (fuelwood shortages) in the rural Global South. Up until this point, energy supply was primarily considered in its implications for

J. Clancy, *Engendering the Energy System*,
https://doi.org/10.1007/978-3-031-78965-6_6

the economy. The Global North caught up when we entered the twenty-first century. The European Commission asked for a study of women in energy in Europe since when more agencies have expressed an interest in gender and energy (see, for example, the International Energy Agency's User Technical Cooperation Programme[1]).

What has emerged from the accumulated body of research is that there is an overlap in the issues faced by women in the Global South and North which opens the possibilities for cross-learning (Clancy 2020). Although women in the South may have more limited capacities to act than their sisters in the North to improve their lived experiences (Williams 2020).

There has been criticism that much of the early work on gender and energy has been carried out by researchers and practitioners without a social science background (see, for example, Listo 2018), which is ironic since it was a gender specialist, Irene Tinker, who could be credited with starting the interest in the gender dimension of energy when she pointed out that the real 'rural energy crisis' is women's time. While I agree that there could be clearer definitions and use of concepts, one can still be a feminist and a scientist or engineer and make a meaningful contribution to gender and energy research. There are also criticisms that too many studies which identify as 'gender and energy' conflagrate 'sex' and 'gender' (see, for example, Fathallah and Pyakurel 2020). One could ask why has it taken so long for gender specialists to discover the energy system? I concede that, unfortunately, there is still a methodological tendency (particularly in quantitative research) to use 'gender as a discrete variable' narrowly construed in binary terms (Özeröl and Harris 2020). However, to move beyond the binary of women and men, requires the availability of disaggregated data which in large scale surveys for government institutions, that many researchers rely on for data, is not the norm. In my opinion, we have come a long way in gender and energy research and practice in the way we now recognise that 'gender' is not another word for 'women' or that it is a binary classification of people into two homogeneous groups called 'women' and 'men'. It is increasingly recognised, at least in research, that women and men differ across a range of social categories (such as age, class, ethnicity, social status, marital status, economic group and sexual identity) which interlink.

[1] https://userstcp.org/gender-energy-task/ (accessed 28 July 2024).

The issues which form the basis of gender and energy have certainly evolved in the last 40 years. Household energy was initially a single issue related to women and fuelwood in the Global South. Elizabeth Cecelski's paper in the influential journal 'Energy Policy' helped to broaden the concept of gender and energy from biomass stoves to one of a broader range of issues including pricing, transport and modern energy forms, such as electricity and LPG (Cecelski 1995). She had already pointed out that the family home could also be a centre of production, particularly for women (Cecelski 1987). Household energy is now increasingly recognised as an issue in the Global North. Household energy is now accepted as more complex than a technical issue of affordable clean stoves and fuels to counteract health issues related to smoke pollution, for example, recognising the embeddedness of a traditional fuel in a society's cultural norms—the size of your fuelwood stack as a measure of being a good mother in South Africa (Matinga 2010) and cutting peat in rural Ireland as an indicator of being a good father (Eakins et al. 2022). Other health issues have emerged such as stress, anxiety and depression arising from having insufficient energy to meet your daily needs (Feenstra and Clancy 2020).

The formation of ENERGIA,[2] the international network on gender and energy, in 1996 created in a significant player in engendering the energy system. It started as a reaction by four women from the Global North to the lack of attention to women and energy in the Beijing Platform for Action. The founders were firmly of the opinion that the network should be run by women from the South since they were best placed to reflect the energy situation as experienced by women in the South, so they gradually reduced their involvement over time while the network grew and has flourished. The network's first activity was to set up a newsletter (ENERGIA News). When looking for material, the editors found that there was rather limited evidence about women and energy particularly compared to other sectors, such as agriculture, and tended to be in the form of small single case studies. However, it tends to be large-scale evidence that motivates policymakers to act. ENERGIA has contributed to building the body of evidence, including with two major research projects: 'Gender as a Key Variable in Energy Interventions' and the 'Gender and Energy Research Programme' which I have drawn on in

[2] Https://energia.org/ (accessed 23 July 2024).

writing this book. ENERGIA has evolved over time, diversifying its activities, including developing training courses to build capacity and carrying out advocacy and lobbying at international, regional and national levels, for which it draws on the evidence it generates to inform policy and practice. An indicator of ENERGIA's international reputation is that it is a co-host of the UN's High Level Political Forum which is a mechanism for monitoring progress with reaching the SDGs.[3]

The gender and energy focus has expanded from rural areas of the Global South to the Global North where attention has been on urban areas. However, that is also changing. It was an assumption that, in the South, energy access was better for those with a low income in urban areas than it would be in rural areas. However, a three-country study (Nigeria, the Philippines and Brazil) about the gender dimensions of energy use in an urban context found that while energy availability might be better than in rural areas, this does not necessarily mean that problems of access had diminished (Clancy et al. 2006). It is also good to see that urban informal settlements, an under-researched area, is now receiving more attention at least from researchers but continues to fall outside of politicians' interests. A significant number of the urban population in the South (for example, approximately 60% in Nairobi, Kenya) live under such precarious conditions which can result in households resorting to unreliable and dangerous illegal connections (Anditi 2022). In the Global North, the cost of mobility, linked in particular to energy prices, is turning attention towards rural areas and public transport vs cars—which have a distinct gender dimension (Lucas et al. 2016; Clancy et al. 2022).

In the Global North, since it was assumed that clean energy availability was not an issue, the focus at the household level was on energy use in which household income is regarded as a significant factor. However, research has shown that it is not only people on low incomes who need help in addressing their energy use but also people with high incomes who have sufficient income to meet their energy bills yet need to be persuaded to lower their energy consumption to contribute to the green energy transition. The closer attention to the social characteristics of the user and their capacity to respond to adopting more energy efficient or green forms of energy has identified a diversity of barriers other than income, such as older people with no smart phone and declining literacy levels.

[3] Https://hlpf.un.org/ (accessed 24 July 2024).

This diversity has led to the emergence of the concept of 'hard-to-reach users' which reinforces the need for a disaggregation of social groups and avoiding 'one-size-fits-all' policy approaches (Feenstra 2021).

Gender and energy research has also contributed to reducing the invisibility of men within the energy system. While responses to the 'fuelwood crisis' had put the focus on women, researchers began to bring men into the picture by providing evidence from a range of communities that it is the male members of the household who decide about household purchases, including stoves (see, for example, Dutta 1997). Such findings contributed to challenging macroeconomists' classification of the household as a unified entity.

While there has been a lot of activity in promoting women as energy entrepreneurs in the Global South (see, for example, ENERGIA's Women's Economic Empowerment Programme[4]), as well as attention to the contribution energy access makes to women's enterprises (see, for example, Clancy et al. 2006), there has been less attention to men's enterprises (see, for example, IDS and GIZ 2019; Fingleton-Smith 2020). Even less attention has been given to energy and women's enterprises in the Global North. The services and commerce sectors, such as food services (restaurants, bakeries, coffee shops) and health and beauty providers (hair salons and nail bars), employ significant numbers of women (around 60% of the work force is the estimate for the EU [European Commission 2022]). Since these businesses are often significant energy users, their viability is vulnerable to energy prices and are often located in rented premises dependent on the willingness of the landlord to improve the energy efficiency of the building. However, in the past SMEs are usually, at least within the EU, excluded from energy consumer protection or tend to be overlooked in advice on energy efficiency improvements. Therefore, the recent policy decision to enable support to 'vulnerable micro-enterprises' is to be welcomed, since the enterprises employing significant numbers of women would fall within that category (Clancy et al. 2022).

When I came to reflect on the evolution of gender and energy research, I realised that while much has changed there is still a lot that remains the same. Women working in the energy sector or as a researcher continue to face the same issues as to whether this is a 'suitable' area for women,

[4] https://energia.org/impact-area/womens-economic-empowerment/ (accessed 26 July 2024).

in other words, conforms to gender norms. The energy sector, globally, continues to have a masculine image. The employee profile is still dominated by men aged 50 and above, particularly in the technical areas while women tend to work in administration, sales, finance, catering and personnel. If a woman passes the selection process, she finds herself in an organisational culture which does not readily proscribe to collaborative ways of working preferred by many women (Moodley et al. 2016) nor an employment policy sympathetic to childcare.

Perhaps one of the most surprising recent findings relates to women accessing finance which they may require to improve the energy efficiency of their business or set up an energy technology company. While it is, unfortunately, not unusual for women to face higher barriers than men in the Global South to accessing finance, it appears that this is a global issue. Data shows that of the capital invested in European technology companies 91% went to all-male founding teams while 7% went to mixed teams and less than 2% to all-female teams (European Commission 2022).

It appears we still have some way to go in engendering energy policy. Again, ironic since one of the earliest studies related to engendering policy was by Barbera Farhar and her colleagues at the United States National Energy Research Centre who set out to determine whether there were gender differences in energy policy preferences in terms of what we now call the energy transition (Farhar et al. 1980). They could find little discernible differences, although women favoured energy conservation slightly more than men did as well as having slightly stronger opposition to nuclear power than men did. The question has continued to be asked and the findings tend to be the same. A four-country study (France, Germany, Norway and the UK) some 40 years later came to the same conclusion about nuclear power except in Germany where there was no discernible difference (Sonnberger et al. 2021). A sign of change in research is that the study also disaggregated their findings across social categories including gender, age, economic status, location and housing tenure.

Gender-aware energy policy frameworks do exist. However, it is the translation of policy into practice where progress is still needed. Most of the implementation relates to meeting practical household needs. I will stress that these are not unimportant issues since they can have implications for physical and mental health. Nevertheless, there are more gender issues related to the energy system, both within and outside the household, that need attention. At the end of Chapter 5, the explanation for the limited progress is a lack of data which, if available may convince energy policymakers (who tend to be men) of the significant contribution to the energy transition to be gained by implementing a gender-responsive energy policy in this chapter. Since women are constrained in access to the paths to influence policy, we will need to take the opportunities which open up by understanding how political processes work. In this context Frenova's work on understanding the participatory processes of women and women organisations within UNFCCC-led climate change governance is particularly instructive (Frenova 2022).

6.3 Don't Assume Energy Access Changes Things: At Least in the Way You Expect

The promised outcomes of adopting a piece of equipment using modern energy or switching to a cleaner energy source do not always materialise either in the time scale imagined or in the way intended. Firstly, change can take time. As reported in Chapter 3, the transition to cleaner fuels in some places, took two decades (see for example ESMAP 1999). Similarly changes in gender roles also take time as women in the North can testify. While access to a sufficient and reliable electricity supply powering household gadgets certainly reduces drudgery, it does not necessarily reduce time poverty or change gender relations. In the UK, over a thirty-year period, there has only been a slight increase in men taking responsibility for their household's laundry (Scott and Clery 2013).

Most of the evidence on time-saving as a result of obtaining equipment driven by modern energy tends to relate to women, with only a few insights into men's activities. A factor in not being able to assess the extent of men's benefits from access to modern energy services in rural areas may be reported as increased agricultural output rather than time and effort saved. Reported time savings tend to be anecdotal rather than measured (ENERGIA 2006). The available evidence from participant observation shows that the use of improved stoves or modern fuels, such as LPG, may not produce significant time-savings in the kitchen—as little as 8 minutes a study in rural India reported (Barnes and Sen 2004).

Access to affordable modern energy services in the household can reduce both time and effort spent particularly by women in reproductive and productive labour which in turn improves well-being and frees up time for leisure and rest. On the other hand, improved energy access does not necessarily lead to time-saving. In China, men were able to migrate to urban areas for better-paid work after the introduction of improved technologies powered by electricity enabled women to take over many of men's agricultural tasks which resulted in an increase in women's workload (Ramani and Heijndermans 2003).

There are positive examples for women of access to modern technology powered by electricity bringing changes to their lives. In rural Kenya there is good mobile phone access which allows women to have access to mobile banking services (University of Oslo et al. 2019). On the other hand, women's workloads have increased because the images they see on television have created new standards for cleanliness (Standal 2008).

It is also not sensible to assume that an organisation which presents itself as a social enterprise will behave in a gender-sensitive, socially inclusive way. Energy communities are organisations in which members share in the production, distribution and use of renewable energy at a local level. They are advocated as a means by which ordinary citizens can both shape and directly benefit from the energy transition. However, evidence shows, at least in Europe, that the membership not to be very socially inclusive consisting of mainly white, middle-class and men with a technical orientation (Lazoroska et al. 2021; Radtke and Ohlhorst 2021).

NGOs involved in development work do not automatically take a gender approach to their work. A study of off-grid supply by NGOs in rural Nepal, Kenya and India (two areas in each country) found four of the eight NGOs took gender-neutral top-down approaches to implementation. There was the 'traditional' job distribution with men filling the technical posts and, if involved at all, women in administrative functions (even women who had engineering qualifications) (University of Oslo et al. 2019).

It is not unusual to read that one of the benefits of time saved due to improved access to energy is that it allows for income-generating activities. However, men are more likely to use any time saved for recreation and leisure, while women tend to use the time for housework and childcare, as well as for resting, socialising and watching TV (Matly 2003). There are some insights into men's reactions to the equipment that new energy sources bring. Men appreciate the improved status they gain due to their clothes not smelling of smoke when their household adopt improved cookstoves (Fingleton-Smith 2020). While in Zanzibar, familes acquiring TV sets has kept men at home in the evenings rather than going out to socialise with male friends (Winther 2008).

An assumption about the outcomes of women participating in income-generating activities is that women will gain respect and status within households and in society, which in turn enables them to contribute to decision-making (Diouf et al. 2020). However, households are heterogeneous complex social constructions and the internal relations and characteristics of adult family members play a role in such changes of gender roles. A study in rural Tanzania found that whether women could influence decisions about investment in their families farms was dependent on the nature of the man and woman's employment (on- or off-farm) and their education levels (van Aelst and Holvoet 2020). Women earning an income does not automatically lead to increased welfare in their household since research found that when women start to contribute to household basic needs, men can decide that they no longer have an obligation to financially support their families (Pueyo et al. 2018).

Generalisations are tempting since they can set aside the need for time- and money-consuming research. However, taking a finding that applies in several locations does not automatically apply in another location. For example, women and men working in the Street Food Sector (SFS) in Senegal and South Africa expressed differences about energy choices. Given that one country is in West Africa and one in Southern Africa, it is tempting to generalise that this is a universal condition, at least for Sub-Saharan Africa, and take appropriate measures without the need for data gathering. However, in Rwanda's SFS, women and men appear to have very similar energy priorities (Mohlakoana et al. 2018). One of the most important findings from the ENERGIA gender and energy research programme, which was carried out in 12 countries in Sub-Saharan Africa and Asia, was: context matters (ENERGIA 2019). Context can make definitive answers naive or irrelevant, in other words local and national research must play a key role in designing interventions that leave no-one behind.

Having a policy statement containing gender-aware text does not necessarily result in policy implementation. We certainly need to understand what causes this policy evaporation. Policymakers globally appear to have a limited understanding of what causes gender inequalities overlooking the structural components as well as entrenched issues such as racism and colonialism (Özeröl and Harris 2020). One obvious solution might be to have more women participating in policy formulation processes. However, certainly in international arenas, this can be reduced to symbolic participation in which women's organisations are recognised as participating non-state actors but policies do not become more gender-responsive. Based on an analysis of meetings related to the United Nations Framework Convention on Climate Change (UNFCCC), the evidence suggests there is both a lack of understanding by the organisers of the gender and social issues and how they can help women's organisations contribute. On the other hand, women's organisations lack the capacity (financial, technical and familiarity with participation processes) to make meaningful contributions (Frenova 2022). To make definitive statements about women's influence is not helped by there being no agreement on how to measure the quality of participation.

Orchestrating behavioural change involves engaging with the target group to present arguments for why change is required. This needs to be done in language that is meaningful to the target group. A study in rural Ireland found that when trying to persuade people to stop using their traditional fuel, peat, the arguments for switching were presented in terms of the contribution to stopping climate change which is at a scale beyond many ordinary people's frame of reference whereas the impact of smoke on their health would be easier to see the consequences (Eakins et al. 2022).

6.4 What, How and When We Measure Counts

What we count and how, as well as when, we count have important implications for outcomes and identification of methods and technologies that contribute to a gender-just energy transition.

Large quantitative surveys are, in principle, a good instrument to provide evidence for policymakers about the presence (or absence) of a gender-just energy intervention, the scale of a problem and the financial costs of any interventions required to solve it. However, if these surveys are entirely quantitative, they may miss opportunities for learning about why a problem exists and how the problem is perceived by the group identified as experiencing the problem. Quantitative data should be supported by qualitative data for which the research methods are considered effective in identifying intangible factors, such as social norms, that influence behaviour and choices about adoption or not of policy intervention (Kooijman et al. 2023). A good example of the revelatory value of qualitative methods comes from the Netherlands in which a study about what influences girls to opt for STEM subjects found that cultural norms influence the perceptions about engineering as a career suitable for girls (Feenstra et al. 2022). So, a programme to encourage girls to take up STEM was not as effective as anticipated because it regarded girls as a homogeneous group and took a 'one-size-fits-all' approach to the messages deployed.

However, qualitative methods also have weaknesses. For example, when an interviewee describes what they do and what they do in practice are not always the same thing. One review of the statements about the activities households did once they acquired electric light estimated that these activities were not possible with the lighting level in respondents' homes (Ramani and Heijndermans 2003). Participant observation can help clarify, in some circumstances, statements from oral reporting. In her study in rural South Africa, Matinga asked women how they coped with the smoke in the kitchen. The most common reply was 'I open the window'. However, in the several months she spent in her study villages, Matinga could not recall seeing an open kitchen window (Matinga 2010).

A danger of quantitative indicators as a measure of gender equality is that they can be reduced to mere box ticking. For example, a specified percentage of women employed in an energy company or a policymaking body does not tell us where they are within the organisation or the extent to which they are able to participate in the decision-making. Nor does increasing the number of women in an energy organisation guarantee gender justice either within an energy organisation or outside in delivering the green transition. Firstly, research has challenged the notion that women are more likely than men to represent women's interests (Childs and Krook 2009 cited in O'Neil and Domingo 2016). If we look at what happens inside organisations aiming to promote gender equity, a common approach is to appoint a 'gender officer' which is usually a woman and so removes any responsibility from men.

There are tensions between academics and development practitioners about the nature of evidence. Time plays a significant part in the way evidence is assembled and what it reveals. Practitioners need evidence about the outcomes of their projects for their final reports. A study which looked at the lessons learnt from mainstreaming in 40 gender and energy projects could find no evidence that the implementing organisations collected any monitoring and evaluation data after the formal ending of the project (Clancy et al. 2016). The implementing organisations are not learning about what works and does not work and the extent to which their intervention reached the predicted outcomes. However, academics can have a much longer time horizon to evaluate the longer-term impacts of an intervention—particularly since change takes time to materialise. So, the challenge is to draw the research findings to the attention of policymakers and implementers, including NGOs and civil society organisations.

Another area of tension between researchers and development practice is a lack of conceptual clarity, for example 'women's empowerment' which tends to be advocated on an instrumentalist definition of improved welfare such as the number of hours saved on drudgery due to accessing labour-saving devices. There is little attention given to addressing the structural factors that hinder women's empowerment with limited analysis of the underlying mechanisms that produce (or limit) observed effects (University of Oslo et al. 2019). Indeed, solutions for improving people's lives should not focus on arguments by technocrats about which technology is best for an end-user but should be about enabling women and men to prioritise their problems, to choose which option meets their needs and fits their circumstances. This is an issue in both the Global South and North—where the technocrats are generally unaware of the social, environmental and micro-economic context in which they are designing or recommending technologies.

The concept of a unitary model of the household conceives of a static homogeneous entity operating in a consensual way and pooling resources misses the reality of many people's lived experience. Households are fluid entities with a dynamic structure, varying in a range of socioeconomic characteristics such as income, class, ethnicity and education, as well as marital status and form (Bell et al. 2015). The cultural setting shapes how community members conceive the concept of a household, for example, it can be a co-residence, a site of joint production or shared consumption, in which members may have kinship links that are intergenerational. Households are also dynamic with the natural life processes of birth, marriage and death changing the composition over time. There are families living across multiple households. In the North, divorce and employment patterns are common causes of this type of fragmentation, while male migration is a contributing factor in parts of the South (e.g. Nepal and India) (MSSRF and CRT Nepal 2019). The household is a place of negotiation, in which women and men define their roles and relations (mediated by informal and formal institutions), exhibiting both conflict and cooperation about the allocation of labour and the distribution of resources which affected individual outcomes.

The acceptance of the unitary model of the household led to data gathering via the head of the household, which in mixed households would usually be a man, meaning intra-household differences are missed. Also, by assuming a consensual model of a household can lead to erroneous conclusions, for example, when aggregate household income rises, it is not a given that all household members' well-being will improve equally (Pueyo et al. 2018). The use in the UK of the concept of 'household reference person' when collecting data is at least a step in the right direction of recognition about the complexity of households (see Chapter 5).

A fundamental question which receives little attention is: who actually sets the research agenda, its priorities and provides the answers? Probably an expert who advises policymakers on their priorities and the nature of the problems they would like addressed or, in the context of the Global South, a well-meaning researcher probably from the Global North. If we look at the topic of stoves, which have been the focus of much of the energy interventions in the Global South, it should be the household family member responsible for cooking who should be identified as the expert. However, it is usually an outsider who is the designated expert who frames the problem decides the research agenda, its priorities and how the results are used. Matinga and Clancy (2020) have commented that the data gathered related to the impact of wood smoke on health appears to remain within formal medical or energy circles and does reach those in rural communities including the 'subjects' of the research and community health workers. The authors have also asked why, in the context of fuelwood use and health issues, very little attention is given to the muscular-skeletal damage women and men can suffer when carrying wood as well as the physical and sexual abuse women suffer while out collecting wood. Is this because combustion emissions can be measured in situ while attributing the causality of skeletal damage is more difficult to ascertain and a reluctance to answer an outsider's questions about sensitive matters provides limited data? This makes the case for ethnographic approaches to data gathering, which despite being time-consuming, are revelatory about lived experiences. An example comes from Denmark which used ethnographic methods to try to understand household decision-making in energy choices. The findings showed that family members when deciding whether to implement an energy renovation to their home, such as a solar panel, take into account not only the financial implications but as to whether the renovation would add to the amount of time they spend on their everyday household tasks (Tjørring 2016).

6.5 Moving Forward

Of course, recommendations from academics usually includes the statement that we need more research. I will not break with that tradition! The body of empirical evidence related to gender and energy remains small and insufficient to make meaningful policy decisions. Far too many policies are formulated to promote a gender-just energy transition based on implicit and explicit assumptions about the characteristics of a situation and what works in a gender-just manner with little attempt to verify these assumptions theoretically or empirically. It is not unusual in project documents related to energy interventions to find realised outcomes and aspirational objectives stated in the same way (my personal favourite is that women will go out at night if there is street lighting). Assumptions are presented as universal truths leading to technological solutions introduced into inappropriate contexts. I have tried in the book to draw attention to some of these assumptions.

We researchers and development practitioners should keep in mind that we are usually outsiders to a community we are aiming to understand and help, so our perceptions of a situation might be different to community members. For example, if you spend twice as much time collecting water for your household as you do in collecting fuelwood, which would be your priority to address? Women can be reluctant to completely abandon fuelwood collection since this can be a group activity involving women of different ages and relationships, which offers opportunities for socialisation and for passing on life's values (Matinga et al. 2013). To the well-meaning engineer from the Global South a solar cooker looks a great piece of simple-to-use technology, using renewable energy, to solve significant problem—health issues related to indoor air pollution. I do wonder if a woman from the Global South would have designed a stove which requires standing cooking in the full heat of equatorial noon outside the house with neighbours able to see what you are cooking. In other words, as outsiders we should always question our assumptions, try to understand the context and ask our target group for their opinion.

Where to go from here? There is still a need for a substantial body of independent evidence to challenge 'gender myths' which have become embedded in the energy system. For those of us engaged in research within the gender and energy nexus, we need to re-examine our approaches. Gender and energy research needs to be more interdisciplinary with greater involvement of social scientists who in turn need to embrace numbers. Researchers can unpack the data and offer explanations about what is going on to help generate a more gender-just green energy transition, in so doing demonstrate the power of qualitative research. We can also make the case for official data produced by governments and international agencies to produced intersectionality. One area where gender and energy research has yet to significantly address is the issues of coloniality and the impact on non-binary gender roles which exist in indigenous communities and how this links to vulnerability of Indigenous peoples (Vinyeta et al. 2015). We are certainly behind our co-researchers in the water sector who have examined how water insecurity is experienced by trans- and gender non-conforming communities (Latchmore et al. 2018).

Another area which requires more attention from academic research are the health issues related to fuelwood use, beyond HAP—which is not to undermine the seriousness of this issue. Effort then needs to be made to ensure that the findings reach those affected, as well as those best positioned to help them (community nurses), in readily accessible ways that relate to everyday practices.

I am going to close the book with a group who need to be brought more into the picture: men! As research subjects in the gender and energy nexus, they are rather missing. However, as researchers and development practice men, from different disciplines have been involved from the beginning.[5] I am interested in what will emerge from when masculinities research examines the energy system. The initial male resistance to gender mainstreaming in energy policy and practice has certainly begun to wane but there is still a way to go. We need male allies and champions, from different communities, so that gender equality in policy and practice is not left as the sole responsibility of women. Men can play an important part in helping create a more gender-friendly and socially inclusive workplace by challenging misogynistic behaviour and drawing attention to discriminatory recruitment and retention practices. Men can also benefit from a gender just energy policy and practice.

[5] For example, Andrew Barnet, Amulya Reddy, Dev Nathan, K V Ramani, Stefan Bouzarovski and Ben Sovacool from research and Luis Sec, Antoine Faye, Doug Barns and Matthew Mendez from practice.

References

Anditi, C.A. (2022). *Co-designing Gendered Energy Innovations for Urban Informal Settlement Households in Kenya. A Case Study of Mathare Valley Informal Settlement*. PhD Thesis, Stellenbosch University, South Africa.

Barnes, D.F. and Sen, M. (2004). *The Impact of Energy on Women's Lives in Rural India*. Washington, DC: ESMAP, The World Bank.

Bell, S., Judson, E., Bulkeley, H., Powells, G., Capova, K. A. and Lynch, D. (2015). Sociality and Electricity in the United Kingdom. *Energy Research and Social Science* 9: 98–106.

Cecelski, E. (1987). Energy and Rural Women's Work: Crisis, Response and Policy Alternatives. *International Labour Review* 126 (I): 41–64.

Cecelski, E. (1995). From Rio to Beijing: Engendering the Energy Debate. *Energy Policy* 23 (6): 561–575.

Childs, S. and Krook, M.L. (2009). Analysing Women's Substantive Representation: From Critical Mass to Critical Actors. *Government and Opposition* 44 (2): 125–145.

Clancy, J. (2020). *Reflection on Engendering the Energy Transition*. In Clancy, J. Őzeröl, G., Mohlakoana, N., Feenstra, M and Sol Cueva, L. (eds.), Engendering the Energy Transition. Palgrave Macmillan.

Clancy, J., Kustova, I., Elkerbout, M. and Michael, I. (2022). *The Gender Dimension and Impact of the Fit for 55 Package. Commissioned Study for European Parliament's Committee on Women's Rights and Gender Equality*. Brussels: Directorate General for Internal Policies: Policy Department C: Citizens' Rights and Constitutional Affairs (Women's Rights & Gender Equality).

Clancy, J.S., Maduka, O. and Lumampao, F. (2006). Sustainable Energy Systems and Urban Poor Livelihoods. In Droege, P. (ed.), *Urban Energy Transition: From Fossil Fuels to Renewable Power*. Elsevier.

Clancy, J., Mohlakoana, N. and Diagne Gueye, Y. (with Muchiri, L. and Shakya, I.) (2016). *Mainstreaming Gender in Energy Sector Practice and Policy: Lessons from the ENERGIA International Network*. Commissioned Study for ENERGIA Gender and Energy Research Programme, ENERGIA, The Hague, pp. 1–74.

Diouf, M., Mohlakoana, N., Sarr, S. and Seydi, B. (2020). *Energy Transition and Gender in the Informal Street Food Sector in Africa*. In Clancy, J., Őzeröl, G., Mohlakoana, N., Feenstra, M. and Sol Cueva, L. (eds.), Engendering the Energy Transition. Basingstoke, UK: Palgrave Macmillan.

Dutta, S. (1997). Role of Women in Rural Energy Programmes: Issues, Problems and Opportunities. *ENERGIA News*, 4: 11–14.

Eakins, J., Power, B., Dunphy, N. and Sirr, G. (2022). *Residential Solid Fuel Use in Ireland and the Transition Away From Solid Fuels*. Report Prepared for the Environmental Protection Agency. ISBN: 978-1-80009-035-4. Wexford, Ireland: Environmental Protection Agency.

ENERGIA. (2006). *From the Millennium Development Goals: Towards a Gender-Sensitive Energy Policy Research and Practice: Empirical Evidence and Case Studies*. Synthesis Report for UK Department for International Development (DFID) KaR Research Project R8346 "Gender as a Key Variable in Energy Interventions". The Hague: ENERGIA/HIVOS.

ENERGIA. (2019). *Gender in the Transition to Sustainable Energy for All: From Evidence to Inclusive Policies*. Synthesis Report of the Evidence Generated by the ENERGIA Gender and Energy Research Programme. The Hague: ENERGIA/HIVOS. https://www.energia.org/new-gender-in-the-transition-to-energy-for-all-from-evidence-to-inclusive-policies/ (accessed 9 March 2022).

ESMAP. (1999). *Household Energy Strategies for Urban India; The Case of Hyderabad*. Report 214/99. Washington DC: The World Bank.

European Commission. (2022). *2022 Report on Gender Equality in the EU*. Luxembourg: Publications Office of the European Union.

Farhar, B.C., Unseld C.T., Vories, R. and Crews, R. (1980). Public Opinion About Energy. *Annual Review of Energy* 5: 141–172.

Fathallah, J. and Pyakurel, P. (2020). Addressing Gender in Energy studies. *Energy Research and Social Science* 65: 101461.

Feenstra, M. (2021). *Case Study Analysis—The Netherlands*. HTR Task Users TCP by IEA: Delft. 19pp. https://doi.org/10.47568/3XR114

Feenstra, M. and Clancy, J. (2020). *A View From the North: Gender and Energy Poverty in the European Union*. In Clancy, J., Őzeröl, G., Mohlakoana, N., Feenstra, M. and Sol Cueva, L. (eds.), Engendering the Energy Transition. Basingstoke, UK: Palgrave Macmillan.

Feenstra, M., Creusen, A., Kottari. M. and van Tulder, F. (2022). *Klimaatverandering en Gender: Aanbevelingen, Beleidsanalyse, Lexicon*. Report (in Dutch) prepared for The Netherlands Ministry of Education, Culture and Science (OCW). Amsterdam: Atria. https://75inq.com/services/library/klimaatveran dering-en-gender/ (accessed 22 July 2022).

Fingleton-Smith, E. (2020). Blinded by the Light: The Need to Nuance Our Expectations of How Modern Energy Will Increase Productivity for the Poor in Kenya. *Energy Research & Social Science* 70: 101731.

Frenova, S. (2022). *Understanding Policy and Institutional Strategies to Support the Participation of Women and Women Organisations Within UNFCCC-Led Climate Change Governance*. PhD Thesis, University of Twente, Enschede, The Netherlands.

IDS and GIZ. (2019). *Unlocking the Benefits of Productive Uses of Energy for Women in Ghana, Tanzania and Myanmar*. Research report RA6, ENERGIA Gender and Energy Research Programme. The Hague: ENERGIA/HIVOS.

Kooijman, A., Clancy, J. and Cloke, J. (2023). Extending Energy Access Assessment: The Added Value of Taking a Gender Perspective. *Energy Research & Social Science* 96: 102923.

Lazoroska, D., Palm, J. and Bergek, A. (2021). Perceptions of Participation and the Role of Gender for the Engagement in Solar Energy Communities in Sweden. *Energy Sustainable Society* 11: 35.

Latchmore, T., Schuster-Wallace, C. J., Longboat, D. R., Dickson-Anderson, S. E. and Majury, A. (2018). Critical Elements for Local Indigenous Water Security in Canada: A Narrative Review. *Journal of Water and Health* 16 (6): 893–903.

Listo, R. (2018). Gender Myths in Energy Poverty Literature: A Critical Discourse Analysis. *Energy Research & Social Science* 38: 9–18.

Lucas, K., Mattioli, G., Verlinghieri, E. and Guzman, A. (2016). Transport Poverty and Its Adverse Social Consequences. *Transport* 169 (6): 353–365.

Matinga, M.N. (2010). *We Grow Up With It: An Ethnographic Study of the Experiences, Perceptions and Responses to the Health Impacts of Energy Acquisition and Use in Rural South Africa*. PhD thesis, University of Twente, Enschede, The Netherlands.

Matinga, M.N., Annegarn, H.J. and Clancy, J. S. (2013). Healthcare Provider Views on the Health Effects of Biomass Fuel Collection and Use in Rural Eastern Cape, South Africa: An Ethnographic Study. *Social Science and Medicine* 97: 192–200.

Matinga, M. and Clancy, J. (2020). *Gender, Firewood and Health: The Potential of Ethnography to Inform Policy and Practice*. In Clancy, J., Őzeröl, G., Mohlakoana, N., Feenstra, M. and Sol Cueva, L. (eds.), Engendering the Energy Transition. Basingstoke, UK: Palgrave Macmillan.

Matly, M. (2003). *Rural Electrification in Indonesia and Sri Lanka: From Social Analysis to Reform of the Power Sector*. ASTAE EnPoGen Project. Washington, DC: The World Bank.

Mohlakoana, N., Knox, A., Ranzanici, A., Diouf, M., Bressers, H., de Groot, J., Pailman, W. and Sanfelice, V. (2018). *Productive Uses of Energy: The Informal Food Sector in South Africa, Rwanda and Senegal*. Research report RA2, ENERGIA Gender and Energy Research Programme. The Hague: ENERGIA/HIVOS.

Moodley, L., Holt, T., Leke, A. and Desvaux, G. (2016). *Women Matter: Africa*. McKinsey & Company.

MSSRF and CRT Nepal. (2019). *The Gender Factor in Political Economy of Energy Sector Dynamics*. Research Report RA3 prepared for the ENERGIA Gender and Energy Research Programme. The Hague: ENERGIA/HIVOS.

O'Neil, T. and Domingo, P. (2016). *Women and Power: Overcoming Barriers to Leadership and Influence*. London: Overseas Development Institute (ODI).

Özeröl, G. and Harris, L. M. (2020). *Gender-Sensitive Analysis of Water Governance: Insights for Engendering Energy Transitions*. In Clancy, J., Őzeröl, G., Mohlakoana, N., Feenstra, M. and Sol Cueva, L. (eds.), Engendering the Energy Transition. Basingstoke, UK: Palgrave Macmillan.

Pueyo, A., Maestre, M., Carreras, M., Bawakyillenuo, S. and Ngoo, G. (2018). *Unlocking the Benefits of Productive Uses of Energy for Women in Ghana, Tanzania and Myanmar*. Report prepared for the ENERGIA Gender and Energy Research Programme. The Hague: ENERGIA/HIVOS.

Ramani, K.V. and Heijndermans, E. (2003). *Energy, Poverty and Gender: A Synthesis*. Washington, DC: The World Bank.

Radtke, J. and Ohlhorst, D. (2021). Community Energy in Germany—Bowling Alone in Elite Clubs? *Utilities Policy* 72: 101269. https://doi.org/10.1016/j.jup.2021.101269

Scott, J. and Clery, E. (2013). Gender Roles. In Park, A., Bryson, C., Clery, E., Curtice, J. and Phillips, M. (eds.), *British Social Attitudes: The 30th Report*. London: NatCen Social Research. www.bsa-30.natcen.ac.uk (accessed 27 July 2017).

Sonnberger, M., Ruddat, M., Arnold, A., Scheer, D., Poortinga, W., Böhm, G., Bertoldo, R., Mays, C., Pidgeon, N., Poumadère M., Steentjes, K. and Tvinnereim, E. (2021). Climate Concerned But Anti-nuclear: Exploring (Dis)approval of Nuclear Energy in Four European Countries. *Energy Research & Social Science* 75: 102008.

Standal, K. (2008). *Giving Light and Hope in Rural Afghanistan: The Impact of Norwegian Church Aid's Barefoot Approach on Women Beneficiaries*. Master's Thesis, University of Oslo, Norway.

Tjørring, L. (2016). We Forgot Half of the Population! The Significance of Gender in Danish Energy Renovation Projects. *Energy Research & Social Science* 22: 115–124.

University of Oslo, TERI, Seacrester Consulting and Dunamai Energy. (2019). *Women's Empowerment and Electricity Access: How Do Grid and Off-grid Systems Enhance or Restrict Gender Equality?* Research report RA1, ENERGIA Gender and Energy Research Programme. The Hague: ENERGIA/HIVOS.

van Aelst, K. and Holvoet, N. (2020). *Bargaining Climate Adaptation Through a Gender Lens: An Inquiry into Decision-Making Processes in Tanzanian Farm Households*. In Clancy, J., Őzeröl, G., Mohlakoana, N., Feenstra, M. and Sol Cueva, L. (eds.), Engendering the Energy Transition. Basingstoke, UK: Palgrave Macmillan.

Vinyeta, K., Powys Whyte, K. and Lynn, K. (2015). *Climate Change Through an Intersectional Lens: Gendered Vulnerability and Resilience in Indigenous Communities in the United States*. Gen. Tech. Rep. PNW-GTR-923. Portland:

U.S. Department of Agriculture, Forest Service, Pacific Northwest Research Station.

Williams, M. (2020). *Reflection on "A View From the North: Gender and Energy Poverty in the European Union"*. In Clancy, J., Őzeröl, G., Mohlakoana, N., Feenstra, M. and Sol Cueva, L. (eds.), Engendering the Energy Transition. Basingstoke, UK: Palgrave Macmillan.

Winther, T. (2008). *The Impact of Electricity. Development, Desires and Dilemmas*. Oxford: Berghahn Books.

Index

J. Clancy, *Engendering the Energy System*,
https://doi.org/10.1007/978-3-031-78965-6

The manufacturer's authorised representative in the EU is Springer Nature Customer Service Centre GmbH, Europaplatz 3, 69115 Heidelberg, Germany. If you have any concerns regarding our products, please contact ProductSafety@springernature.com

Printed and bound by CPI Group (UK) Ltd, Croydon, CR0 4YY
05/10/2026
02228125-0006